T0100037

SUPERCITIES ON, UNDER, AND BEYOND THE EARTH

SUPERCITIES ON, UNDER, AND BEYOND THE EARTH

Housing, Feeding, Powering, and Transporting the Urban Crowds of the Future

Jeff Dondero

ROWMAN & LITTLEFIELD
Lanham • Boulder • New York • London

Published by Rowman & Littlefield
An imprint of The Rowman & Littlefield Publishing Group, Inc.
4501 Forbes Boulevard, Suite 200, Lanham, Maryland 20706
https://rowman.com

6 Tinworth Street, London SE11 5AL, United Kingdom

Copyright © 2020 by The Rowman & Littlefield Publishing Group, Inc.

All rights reserved. No part of this book may be reproduced in any form or by any electronic or mechanical means, including information storage and retrieval systems, without written permission from the publisher, except by a reviewer who may quote passages in a review.

British Library Cataloguing in Publication Information Available

Library of Congress Cataloging-in-Publication Data
Names: Dondero, Jeff, 1947– author.
Title: Supercities on, under, and beyond the earth : housing, feeding, powering, and transporting the urban crowds of the future / Jeff Dondero.
Description: Lanham : Rowman & Littlefield, [2020] | Includes bibliographical references and index.
Identifiers: LCCN 2019042357 (print) | LCCN 2019042358 (ebook) | ISBN 9781538126714 (hardcover ; alk. paper) | ISBN 9781538126721 (epub)
Subjects: LCSH: Cities and towns—Forecasting. | Urban transportation—Forecasting. | Smart cities—Forecasting.
Classification: LCC HT151 .D56 2020 (print) | LCC HT151 (ebook) | DDC 307.76—dc23
LC record available at https://lccn.loc.gov/2019042357
LC ebook record available at https://lccn.loc.gov/2019042358

♾ ™ The paper used in this publication meets the minimum requirements of American National Standard for Information Sciences Permanence of Paper for Printed Library Materials, ANSI/NISO Z39.48-1992.

CONTENTS

PREFACE

In the early 21st century, perhaps the most important artistic genre is science fiction. . . . [It shapes] how people understand the most important technological, social, and economic developments of our time.

—Yuval Noah Harari, *21 Lessons for the 21st Century*

This fanciful and factual peek into the future of cities is as much scientific possibility as it is fiction, but the flights of the imagination herein are based on real science. Some of these castles in the sky admittedly may not come to fruition, at least not in the next several decades, but most will be at least in the planning stage by the time we're securely on the moon or Mars. They are based on the projections and perceptions of some very sage and sapient minds.

As far as colonizing other vistas is concerned, I think humankind has proven beyond a shadow of a doubt that any challenge that lies over the next horizon, whether it is on, under, over, or above the earth, will not be left unmet. And heading for the stars may be the biggest hurdle humans have faced since they dropped down out of the trees.

This book may read like a combination of sci-fi tech-talk. In a way I guess that makes me a futurist. But this book is meant to provide a glance into the next generations of food, building materials, structures, modes of transport, and possible lifestyles. It is also a wake-up call to stop and think before we get swept up in the seductive vortex of technology, and to reflect on the fact that the future does not just happen; it is created by us, and we will be held responsible for the decisions today we make for tomorrow. We are now taking steps not by mathematical projection (one plus one) but in a logarithmic manner (ten times ten), and these are the biggest techno-strides we will have taken in human-kind's history.

One thing is for certain: there is a groundswell of people thronging back into our cities, and our metro spaces are not prepared for it. What we find in cities today are exorbitantly expensive houses, buck-busting rents, deteriorating inner-city transportation, embarrassing homeless situations, inadequate treatment of waste and pollution, and the need for improvements everywhere from roads to resources. Our cities are dealing with the chasm of difference among the socioeconomic classes in terms of distribution of the necessities of life; restrictive rules, codes, and regulations in building; bureaucratic bungling; and poor planning.

There have got to be some changes made—and not old, passé Band-Aid fixes but far-reaching, novel ways that not only combat our current problems but also provide sustainable innovations that will work for the ages, not just for a few years.

We are living in a cross between yesterday's science fiction and soon-to-be scientific fact; for what is fantasy today may well be reality tomorrow. It will be run by technology, which, like a force of nature, nothing can stop and which we have only just begun to understand and respond to. In the future our smart houses and household robots loaded with artificial intelligence may be running us instead of the other way around. We are at the gateway to new and incredible products, materials, and technology, but how are we going to go about managing or manipulating them for our benefit?

The present mindset of our culture is often driven by incompetent politicians, scarcity-oriented economics, and a system of passé values that have to be reevaluated in the face of major disruptive changes. In order for us to be able to make the transition to this new age, quantum leaps in both thought and action are required. Experience tells us that human behavior can be modified toward and through constructive or destructive activities.

It looks like we take the best and worst of us wherever we go; despite our raised consciousness about husbanding resources, protecting the planet and its contents, we also are littering our nearest alien celestial bodies, forming a nascent "space force" despite a treaty that promises we will not do so, and creating an atmosphere of flag planting and raw materials hunting. It has been the hallmark history of humankind to put profit and possessions before people and our planet.

It seems that too often we forgo real personal contact with other people, friends and family alike, in favor of time spent with our electronic gadgets. I have a friend who has a cabin in the woods that has no connection to anything modern and electronic—no TV, cell reception, Wi-Fi, or internet server. Consequently, people in the cabin actually engage in conversation, take walks, and do things together—while looking straight ahead or at each other, not downward into devices.

One thing that we should not ignore, going forward, is inner-city interaction with "green"—it may be the most important color in the city. Many studies have concluded that plants, parks, and quiet natural areas are absolutely essential for relaxation, critical thinking, and physical as well as mental well-being.

One of the ways technology has bloomed is with building materials and methods. Their many forms and uses are nothing short of spectacular. Even old foes like carbon dioxide are going to be either cut back, sequestered, removed, or made into building materials, and recycling and landfill mining will be major megaindustries.

Boomers growing up thought the ultimate technology was the A-bomb and that it was going to bring about our apocalypse. Now technology will be one of our saviors. We now look at science as on the cusp of being divine—some even ask the question of whether God is technology or technology God.

Ironically, maybe we spend too much time isolating ourselves, disconnecting from the real world around us in order to maintain the perception of being connected to the world of today and tomorrow with apps, bots, programs, and computers in "the cloud."

Even though we are heading toward disruptive and major changes, I am confident that we will not only survive but thrive. Human beings are, if nothing else, essentially problem solvers and progress makers. And we all know that we have lots of problems to solve and much progress to accomplish. But how, when, and in what ways all will be consummated is, well, a crapshoot.

ACKNOWLEDGMENTS AND DISCLAIMER

My thanks go to the many people who assisted in the creation of this book, people whom I call curators of information. They tirelessly gathered facts and figures to illuminate the world in order that others may ponder it.

As always, special thanks to my sounding board Patrick Totty for advice, research, suggestions, encouragement, and conversation over a few cocktails.

To my online editor, Suzanne Staszak-Silva, and my production editor, who have been patient and understanding in getting this manuscript into publishing form, and to Rowman & Littlefield for giving me the opportunity to write for others to read.

DISCLAIMER

According to whoever said it first, "There are lies, damn lies, and statistics." Most of the stats here have been gathered by local, state, federal, and international government agencies; scientists; writers; power and utility companies; universities; other informed sources; and the Oz of the internet. Consequently, I do not claim that all the statistics presented represent accurate and true statements, percentages, and facts, and I do not warrant or make any representations as to the content, accuracy, or completeness of the information, text, graphics, charts, web links, websites, and other items contained in their media presentations.

Aggregating and writing information for this kind of book has its inherent problems and predicaments. When presented with the same questions, different people use different ways to find diverse answers and conclusions. Consequently, answers may vary, sometimes quite a bit. In addition, some of the facts presented may be affected by time, changing world events, or new discoveries. As in many things, opinions vary as to number, percentages, predictions, and the veracity, or divergence from same, of the results obtained by individuals using the same information. No one can completely and accurately predict the future. And not all scientists would agree on matters such as global warming and climate change due to greenhouse gases, space exploration and colonization, or the availability of certain products and methods listed herein.

What I have tried to do is present an informed opinion about how homes and metropolises will be built, look, and function. I have also stated that there have been doomsayers that made and continue to make dire predictions about the fate of the earth—I hope I'm not viewed as one of those people, but I figure it's best to err on the side of caution and not carelessness.

Although most of the facts presented herein are defensible, I use them as literary, entertainment, and educational devices to give the reader a general perspective on the subject of resources, energy, materials, the waste we create, and how we may live and/or cope with them in the future. In an effort to communicate more easily and effectively I have taken some averages, mean numbers, and common sense and have modified statements in order to reflect more than one set of opinions and/or updated them by logic and informed opinion.

The lawyers put it another way: Neither the author or publisher, nor any of their employees, makes any warranty, or guarantee, express or implied, or assumes any legal liability or responsibility for the accuracy, completeness, or usefulness of any information, percentage, apparatus, product, device, or process disclosed, or represents that its use would not infringe privately owned rights. Reference herein to any specific commercial copyright, product, process, or service by trade name, trademark, manufacturer, or otherwise, does not necessarily constitute or imply its endorsement or recommendation. The author does not receive any recompense or trade-offs for any product or persons mentioned in this book.

The views and opinions of authors expressed herein do not necessarily state or reflect those of Rowman & Littlefield Publishing Group or its agents. This book is for entertainment and educational purposes only.

I

SEEKING SUPERCITIES

Surroundings, Settings, and Situations

A city is not gauged by its length and width, but by the broadness of its vision and the height of its dreams.
——Herb Caen, columnist and author

Where trade routes met or a body of water was used as a port or a mother lode of something valuable was found, cities inevitably blossomed nearby. Throughout history, cities have been the heart of the social, cultural, and economic development that has made up much of the fabric of civilization. With cities came divisions of labor, which gave people the time to create cultures and which freed them from the grim drudgery of subsistence living.

And as cities prospered, they grew in size as people moved from farms to factories. But just because they were big doesn't mean they were the best places in which to reside. In fact, a large population base has been both a hindrance and a help when figuring out whether a city is a super place to live.

As a fuel source coal provided a cheap and efficient source of energy for steam engines, furnaces, forges, and homes across the country. It spurred massive economic growth and was considered a boon for cities. Not until recently have we seen the kickback for the overuse of fossil fuels.

Buildings and urban infrastructure account for 40 percent of raw-material use, one-third of energy consumption, and 70 percent of greenhouse gas emissions, in addition to having ecological footprints several hundred times larger than their actual acreage, according to UN-Habitat.[1]

Problems that are already visible today—land and food shortages, heavy traffic loads, urban and global warming, congestion, and air pollution, to name a few—will worsen in the future. Consequently, scientists and other practitioners have been looking into measures to address these challenges, especially when reenvisioning old and making innovative plans for new buildings. Creators of supercities will have to find answers for how to supply the needs of megapopulations and manage metros of twenty to one hundred million people and all their trappings.[2]

A LOOK BACK

In 1800 about one out of every twenty Americans lived in cities.[3] In 1860 no city in the United States had a million inhabitants.[4] By 1890 New York, Chicago, and Philadelphia had passed the million mark, and by 1900 New York had 3.5 million people (then the second-largest city in the world).[5]

Around 1790, spurred by the onset of the Industrial Revolution, country folks started to feel the pull of the city, because new and better-paying jobs were far superior to farm labor and subsistence living. At the dawn of the 1800s, on average only about one out of every twenty Americans lived in cities, but one hundred years later many cities had grown by about 35 percent.[6] Cities were becoming fashionable in addition to offering preferable employment.

Between 1880 and 1890 almost 40 percent of the townships in the country lost population because of migration to cities.[7] The development of multistory buildings and public transportation made it easier for people to find places to live that were more accessible to their employment.

Since then the population of cities has more than doubled. People were drawn to cities because of employment opportunities that made their labor more valuable due to division of labor. When on the farm, people had to work for everything. They grew their own food, made their own clothes with materials garnered from animals they raised, built their own shelters, had only a modest choice of goods—nothing as exotic as wine or spices—and had little time for recreation. It was a hand-to-mouth existence.[8]

At the turn of the nineteenth century, about three out of ten people populated US cities. The proportion changed dramatically by 1920 to about one out of two, then two out of three in the 1960s. By the end of this century, about 80 to 90 percent of US residents will live in cities.[9]

After WWII the "American Dream" was life outside the city, complete with a car, white picket fence, 2.4 children, 1.0 dog or cat, a little green grass, a backyard barbeque, and maybe even a pool.[10] Fleeing the hordes in the city, "suburbanites" coveted their single-family dwellings, improved schools and parks, shopping centers, commuting in their own cars, drive-through living, and a sense of the peace, privacy, and comfort living in the "country." They wanted upward mobility and clean, safe places to live. The luxuries of suburbia beckoned—but not for everyone.

Until the Fair Housing Act was enacted in 1968, government-sponsored home loans could be granted based on the color of one's skin—Caucasian was "get-ahead green" and red was everything else. Fully 98 percent of home loans were granted to white folks; the process of excluding all others was called redlining.[11] Ever wonder why the majority of people living in suburbia look the same even now? According to Nikole Hannah-Jones, a reporter for the *New York Times* specializing in civil property–related racism, even though redlining is illegal, every year four million people of color face rejection of their loan and insurance applications.[12]

BACK TO THE FUTURE

At the pace humankind is reproducing, we will reach a population of 9.7 billion within the next thirty years. Every day another 250,000 humans are born.[13] The population of the United States will pass a half billion by 2050, and when the earliest baby boomers reach the age of eighty, they will have witnessed the population of the world triple.

Within thirty-five years more than one hundred cities will have populations larger than 5.5 million people,[14] including twenty-seven supercities with ten million and close to twenty megacities with approximately thirty million inhabitants.

This puts an incredible strain on dwindling resources. The effects of overpopulation are becoming more radical: increasing global surface temperatures, depletion and pollution of the biosphere's resources, waste of water, species extinction, and deforestation. These conditions have been around for many years, but their growth is becoming more swift and alarming.

We have to think about and plan for not only where we are going to put all these people but also how to do it economically, swiftly, sustainably, and humanely. At present the world's cities occupy only about 2 percent of the earth's surface but house almost 60 percent of its population.[15] And as cities grow, they also seem to become outrageously expensive; the two most expensive places to live in the United States are San Francisco and New York.

According to the US Census Bureau, 80 percent of the population were already living in urban areas in 2010.[16] More than 95 percent of the country's most populous state, California, live in metro areas,[17] and that number is projected to grow. Many cities are building "up" because they have no room in any other direction. San Francisco and New York are good examples. The total area covered by the world's cities is set to triple in the next forty years. This will make inner-city property the next to be gentrified.[18] This rush to cities is exacerbating already monumental problems of traffic congestion, air pollution, lack of dependable power, lack of public transportation, larger parking lots, and flawed overbuilding.

Over the years a subtle shift had taken place. Younger people seeking more affordable, more fashionable, or newer housing were moving from older suburbs to exurbs—rural areas surrounding suburbs—resulting in the decline of the suburban nation.

THE GREAT INVERSION OF THE MOVABLE MILLENNIALS

In the past couple of decades, there's been a "great inversion."[19] For the first time in nearly a hundred years, the rate of urban population growth has outpaced suburban growth. This is going to cause problems in that cities are already crowded, expensive, sometimes dangerous, and, with ever-decreasing land to build on, lacking places to put people. We are long overdue for beginning to plan for contemporary megametro locations to ensure they're healthy, self-contained, sustainable, and economic to build, operate, and occupy.

The children of baby boomers have eschewed the lifestyle of backyards, burbs, and barbeques. Millennials haven't experienced a baby boom of their own and are also delaying the launching of one. At present, the nation's birthrate is going down, and there are more baby boomers and seniors in many suburbs than there are families with young children.[20] The taxpayer base in suburbs is increasingly made up of older folks, as millennials are choosing to settle in urban areas, leading to the decay of suburbia. Aging boomers don't care about schools, more parks, or recreation; they want the support services they'll need as they age.[21]

Millennials and Gen Xers are not buying single-family dwellings; instead, they're renting, and some are still living with their parents or grandparents, in part because of hefty student debt, tight mortgage-lending standards, and the heavy buy-in price and extra cost of the traditional suburban lifestyle. Homeownership levels among heads of households thirty-five years or younger was at 36 percent in 2015, the lowest figure since the Department of Commerce started tracking that data quarterly in 1994.[22]

Millennials no longer desire—nor can they afford to buy—super-sized suburban McMansions (homes built between 2001 and 2007 and having between three thousand and five thousand square feet of space).[23] Even more modest homes are being priced out of reach. Construction of single-family homes fell by about one-third between 2005 and 2015, and construction of apartments and condos is at the highest level in forty years.[24] Malls used to be a big draw in the suburbs, but now anchor stores like Macy's, Sears, and JCPenney are closing by the hundreds, and other chains are moving from suburban areas back to cities.

Millennials are getting older, and studies show they want to live where they can walk for recreation, services, and shopping—whether that's in or outside of a city. More than 60 percent of millennials have chosen to rent over buying a home.[25] They are the country's biggest migrators, representing 43 percent of the United States' most restive population, despite making up only 23 percent of the total population.[26]

Millennials and even some baby boomers are ditching the suburbs for major metros everywhere. And regardless of age, urban dwellers see eye to eye on their vision for the future. They want city life punctuated by parks and playgrounds, an increased ability to bike or walk around their neighborhoods, but also a "bright lights, big city" atmosphere.

It behooves us to remind ourselves of the potential population bomb—humans aren't just prolific and the apex species on earth. Humans make up the only species capable of radically expanding its population, changing the face of the earth; the singular species that can increase its life span, and force other species into extinction through pollution, tampering with the environment, greed and appetite; and the only species capable of causing its own demise in more ways than one.

URBS VERSUS BURBS

The question is not really which is fading but how things are changing. In gentrified areas the best and most expensive real estate—whether in a city or a town—is occupied by the well-to-do, either from being there first or by seeking out the best parts of a city to gentrify. But whether their inhabitants are wealthy or not, cities of tomorrow have to be elderly-friendly, color-blind, and age-neutral.

Cities and suburbs alike started to suffer when major manufacturing markets pulled up stakes and moved because of economic conditions, costing people their jobs. Property values plummeted in places where manufacturing was key, and the cities fell into decay. Suburbs started to languish because young people either couldn't afford the price tag or didn't care for their parents' lifestyle. Tax bases began to stagnate, and "anchor stores" in towns and malls started to shut their doors.

In the suburbs cars became a necessity for getting anywhere—one result of suburban sprawl and some pollution. Goods and services in low-density neighborhoods are farther away, and walking to get anywhere can be difficult. Many places in the burbs are unsafe due to traffic, especially for children. Autos create four times the carbon footprint of higher-density neighborhoods and require roads, parking, and auto support systems. Chauffeuring is required to and from children's playdates and activities, as well as for doctor's appointments, requiring more time and miles logged by those who are already driving to work. This is also increasingly a problem for boomers who are partially responsible for their parents' financial, physical, and emotional care as well.[27]

The compulsory commute junket is expensive, for both one's wallet and well-being. According to Ivica Marc, a personal trainer at Exceed Physical Culture in New York City, "If you are sitting in a car, train, or bus for long periods of time every workday, you are putting yourself at risk for heart disease, diabetes, and premature death."[28] And commuting to and from work harms our psychological health and social lives; it can be even more exhausting than the work itself.[29]

Research has been mounting that establishes a link between the sprawl of our living spaces and the rise of obesity, blood sugar, blood pressure, body weight, and metabolic risks—even a rise in divorce.[30] In the burbs, even the fairly useless grass is imported, fertilized, doused with herbicides, and protected by neighborhood landscaping codes in many places.

When one's kids are young, living outside the city might feel safer, but when one's parents get older, suburbia can become a prison, because older folks need to be driven everywhere and looked after. It may turn out that staying in the burbs will be less healthy or safe and provide less opportunity for independence.

A disturbing trend is the demise of the mainstay sport of suburbia: golf. More than eight hundred courses have closed their clubhouse doors in the past decade. The Sports and Fitness Industry Association claims that millennials between the ages of eighteen and thirty agree with Mark Twain, who supposedly said that golf is a good walk ruined, or they just can't afford the previous generation's country club lifestyle.[31]

In high-density cities you can walk just about everywhere. City residents now prefer to drive a mile or two instead of ten or twenty, and they own one car instead of two. The mantra "Location, location, location" is being replaced by "Access, access, access." Cities are starting to go country as they value walking and biking, green surroundings, contact with cultural interests, and living within their means. Urban planners are taking notice all over the country. It may seem counterintuitive, but in denser cities ten times more tax money per acre is generated than by their country cousins.[32]

URBAN SPRAWL

Like middle-age spread, urban bottoming-out is due to poor planning, overexploitation of resources, greed building over green building, poor public transportation, and overreliance on cars.

City populations have suffered from a concentration of inequalities, including poor housing, low-quality education, unemployment, and difficulty or inability to access certain public services, such as health care, welfare, and green spaces, in addition to the decay of high-density neighborhoods into ghettos.

Making cities green and healthy for everyone goes far beyond simply reducing greenhouse gases and planting turf or a tree here and there. A holistic and healthy approach to the environment and resources has to be adopted. For example, suburban redevelopment and rapid transit in Atlanta were planned around downtown areas, including twenty-two miles of parks and developments, in a loop around—not through—city neighborhoods.[33]

According to John Wilmoth, director of the United Nations Department of Economic and Social Affairs, Population Division, "Managing urban areas has become one of the most important development challenges of the 21st century. Our success or failure in building sustainable cities will be a major factor in the success of a United Nations development agenda."[34] Neighborhoods have to be inclusive and open, protected by proactive choices, making funding more equitable and ensuring that no neighborhood is forgotten in urban planning.

GOING UP/VERTICAL SPRAWL

In New York City the record-breaking 432 Park Avenue is taking housing to new heights—1,396 feet to be exact, standing as the tallest residential building in the world and the second-tallest building in New
York City.[35] It's also taking shots from detractors who are saying that it's
too skinny and too rich. In fact, it has come to light that many of the
most luxurious residential projects are also conspicuous consumers of
energy and that they create a chasm of lost sunlight in the "street can-
yons" below.

Many people are suggesting that urban building and zoning codes
have to be changed or at least relaxed. Some urban planners want to
bring back a disappearing concept called "the missing middle," complexes of small condos or individual units with shared outdoor space.
It's the happy medium between a single-family, detached home and a
ten-plus–unit apartment. Think of them as more practical urban models
of tiny homes, which are becoming chic for some folks but unrealistic
for a lot of lifestyles and city codes.

City planners, designers, and forward-looking thinkers are pushing
the limits of creative thinking to envision future cities of all types. From
super-skyscrapers soaring many thousands of feet upward, cities floating on or under water, burrowing underground, in orbit, or creating
skylines on other planets—the only limit is imagination. Cities of the
future will include

- floating sea cities
- high-rise or rooftop farms
- 3D-printed homes
- buildings with their own microclimates
- huge bridges that span entire cities
- spaceports with easy access to the moon and Mars
- superhigh buildings
- underwater and underground cities
- collapsible and stackable living pods[36]

GETTING AROUND

Getting around in the United States is becoming a huge hassle. Our
roads, highways, and bridges are desperately in need of repair. The
American Society of Civil Engineers (ASCE) gives the United States a
D for its roads and a C for its bridges—which is generous grading.[37]

The US Department of Transportation estimates that almost $1 trillion is needed to revamp the current interstate and highway system in the country.[38] Unfortunately, there won't be many highways improved or bridges repaired because we can't afford to maintain what we already have. And research shows that reducing highway congestion by adding more lanes—a phenomenon called "induced demand"—is counterproductive as it ultimately just adds more wheels on the pavement.

"Traffic jams are getting worse, queues longer and transport networks more prone to delays, power outages more common." The United States is a backward country when it comes to passenger trains. As anyone who has visited Europe, Japan, or Shanghai knows, trains that travel at 200+ mph have become everyday modes of transportation.[39]

It's time we design a future where driverless cars, aerodrones, and new-age subways zip around, under, or over skyscrapers, and vertical gardens are in hyperconnected, energy-efficient "smart cities." The alternative is being trapped in endless traffic jams while infrastructure crumbles and pollution overwhelms the remaining declining green spaces.

CARMAGEDDON

Several cities are starting to ban one of our most cherished personal possessions—the car. And it may be one of the safest and healthiest things to do. In 1900 nobody was killed by cars in the United States because they were few and far between. Just twenty years later, as Peter Norton, a professor at the University of Virginia, wrote in his book *Fighting Traffic*, more than two hundred thousand people were killed by cars. In 1925 alone, cars killed about six thousand children.[40] And with small, self-driving electric vehicles (EVs), skyports, drone delivery service, and mass transit, there will be less need for autos and trucks, parking, driveways, and pavement and more room for playgrounds, parks, and housing. Cities have become far too car-centric; autos and trucks are far too much a part of people's lifestyle; and vehicles make city walking more like a bullfight than a stroll. The car, Norton points out, is the lowest-density means of transportation—and most expensive mode of transportation. In the United States more than 90 percent of all trips were taken by car; too few people are moved in a single vehicle and too much fossil fuel is used moving them.[41]

Walk Friendly Communities is a national recognition program developed to encourage towns and cities across the country to make safer walking environments, and in 2011 the Pedestrian and Bicycle Information Center announced the selection of eleven Walk Friendly Communities across the country.[42] Cities such as Los Angeles (no kidding) and Seattle aim to reduce parking spaces and convert some roads and bridges for use by pedestrians and bikes. A good model for them may be Las Ramblas in Barcelona, Spain—a tree-lined thoroughfare.

THE DIFFERENCE BETWEEN A MEGACITY AND A SUPERCITY

A megacity generally has a population of around ten million. In the future, supercities might top out at one hundred million or more.[43] It's been estimated that by 2100 we can anticipate cities of 140 million people—picture Tokyo, Mexico City, New York, São Paulo, Mumbai, New Delhi, and Shanghai all rolled into one.[44]

A supercity can be humankind's dream place. It is a self-contained, quality living organism where there is enough affordable housing that is close to healthy outdoors experiences; the air is refreshing; the water is pure; energy is clean, ample, and inexpensive; food is grown, raised, or produced within one hundred miles of the consumer; building materials are recycled sustainable, and green; transportation is easy and affordable; and all the necessities and luxuries of life are nearby and do not require an oil tycoon's bank account.

The problems and challenges supercities will face include the issues of carbon-neutral environments, how to control urban sprawl and traffic congestion, how to solve the predicament of the homeless, and how to organize, be administered, and be operated sustainably. Adoption of widescale use of renewable energy will be embraced, waste management will become a major industry, and biodiversity will enhance the natural environment. Green transport systems, innovative materials, and construction methods will be utilized; and a diverse population will enjoy a healthy outdoor environment. Some of these enhancements are already being applied or experimented with today, and some are in various stages of planning for use tomorrow.

As more people flood cities, straining already bulging budgets, stretching resources, and staggering city services, some suggestions and solutions will sound reasonable and practical, while others will be the stuff of both science and fiction.

CHRYSALIS CITIES

San Francisco and Manhattan are good examples of city price-out and bound-up boundaries—they have no room to build, save straight up. Following the collapse of the housing market in 2007, the median price of a home was around $700,000 in San Francisco, compared to today's whopping $1.25 million plus, and some rents increased in 88 percent of the nation's biggest cities.[45] When a city begins to run out of horizontal space, it only has a few choices—go vertical, go radical, or go to hell.

Middle-class employees such as teachers, office workers, city workers, and retail workers have been priced out of cities. And don't look to the suburbs for relief—they are bursting their borders too. In places where there were pasturelands a couple of years ago, there are now hundreds of houses, condos, and apartment buildings. And the cost, whether buying or renting, is an unbelievable and unrelenting upward arrow, creating a population of "haves"—and saying to hell with the "have-nots."

THE "GREEN" FOOTPRINT

Creating or maintaining a city's "greenprint" is a tough goal while controlling a rapidly expanding city. And while managing sustainable energy, water, and waste, as well as fostering sensible growth, leaders must keep a green city economically viable and sustainable for the long haul.

Cities of the future will strive to be carbon-neutral—that means they will run entirely on renewable energy, with little or no carbon footprint. Masdar, in Abu Dhabi, might be the first carbon-neutral city in the world, which is ironic since it is being built through the sale of the country's rich oil reserves. At present no carbon-free cities are being planned in the United States.[46]

As we move farther away from the natural world, contact with it becomes more valuable: Urban designers now recognize that access to green space is an important part of people's quality of life. From New York to Singapore, the world's great cities are now placing heightened importance on new and existing green spaces with sustainable urban planning, with the hope of protecting their futures, for both physiological and psychological well-being.[47]

VERY SMART STUFF

The answer to the question of how smart building can be may lie with big data and the so-called internet of things (IoT), where objects previously dumb are made smart by being connected to one other. One way to accomplish this is to plant sensors throughout a city's lands (or in its buildings) to make up a city dashboard, which takes the pulse of the city. This will allow multiple systems to be joined and ultimately work more efficiently to monitor everything from energy use to water and waste, city temperature, traffic patterns, and security. When systems do not "talk" to one another, they operate in isolation, and facility staff are unable to get a holistic view of building performance. This is one of the reasons why building energy management systems (BEMS) emerged to integrate a multitude of disparate systems and functions.[48]

There are those who will loudly howl or silently grumble that this is nothing less than a precursor to *Brave New World*, or *1984*, where "Big Brother" is looking over, under, and around your shoulder. This is partly true. But it is a compromise of privacy in exchange for safety. (See chapter 8, "Getting Somewhere from Someplace.")

ENERGY THAT KEEPS ON GIVING—WITHOUT LEAVING A MESS

Fossil fuels still represent over 80 percent of total energy supplies in the world today.[49] But extensive use of alternative energy sources will allow cities to eventually achieve carbon neutrality. And with the advent of modular smaller grids, not only will power outages be eliminated but an excess of energy will also be left, to be saved or shared on other small electrical grids, which can be connected to larger county, state, or national grids to help create, save, and distribute energy where it might be needed.

Net Zero

One of the first options that should be considered is establishing net zero carbon dioxide–clean cities. Only a couple of these are being built in the world today, and none is being planned in the United States.

California's Energy Commission unanimously passed a law mandating that all new residential buildings up to three stories tall must be equipped with solar panels by 2020, making California the first state in the nation to mandate access to solar in virtually every new home. In addition, the new provisions include a push to increase battery storage and reliance on electricity over natural gas. This is seen as a key measure to decarbonize the building sector—an area that, when electricity use is factored in, represents the second-largest source of greenhouse gases in the state. The rule will likely eclipse the state's current energy-efficiency goal, approved in 2007, requiring all new homes to be zero net energy users by 2020, which regulators now say is not enough to offset a building's use of fossil fuel–derived electricity at night.[50]

Battery storage in homes and businesses (and electric cars) will eventually serve as a giant electricity bank for renewable resources. In this scenario, known as "partial grid defection," homeowners would generate and store 80 to 90 percent of their electricity on-site and use the grid only as a backup—transforming buildings into small power plants and minigrids.[51] (See chapter 9, "Priorities for Power.")

WATER, WATER FROM EVERYWHERE

Water, thankfully, will probably not be an issue in the future. We get water from lakes, rivers, and underground sources such as aquifers. These, along with water saved in cisterns, water desalinized from oceans, and water treated from toilets to tap, are being used for drinking water. This will be accomplished by using state-of-the-art treatment technologies powered by solar energy. Potable water will be stored, ready to use in buildings. Used water will be cleaned and filtered underground via something much like mini–electrical grids, in cisterns or aquifers, where the water will be stored and ready to be used again and again.

COMESTIBLES FOR THE CITY

Concerns that are already being addressed in some cities are the food deserts (where there are fewer places to get fresh and healthy groceries but where there are plenty of fast food and liquor stores) and the rising demand for fresh food from farm to fork. One of the solutions is "vertical farming," or "agri-tecture," which is based on farming that grows upward, around, in, or on buildings and can produce more than enough for residents. Today's largest vertical farm is located in Michigan and is home to seventeen million plants.[52] Other types of soilless farming include hydroponics, aeroponics, and aquaponics, which produce fruits and vegetables, fish, and ducks simultaneously. (See chapter 10, "Provisioning the Populace.")

ULTRAMODERN MATERIALS AND CHANGING CONSTRUCTION

As governments look for ways to adopt green construction codes, they will put more pressure on the construction industry to change the way buildings are designed, constructed, operated, and dissembled. Bleeding-edge materials, innovative uses of old materials, and various applications for recycled materials are nothing short of mind-stretching.

Thanks to recent advances in robotics, computing, and other technologies, a growing number of scientists and engineers think robot-made housing might finally be possible soon. Robotic construction will increase the speed of construction, improve its quality, and lower its price. (See chapter 7, "Bleeding-Edge Building Supplies.")

VACANT AND ABANDONED: A NEW URBAN RENAISSANCE

Buildings that are abandoned and that have physically deteriorated are another vexing problem that older cities face and a plague in all parts of the country. As a quarter of Detroit's population drained out of the city from 2000 to 2010, tens of thousands of buildings became hazards instead of homes.[53] A survey examining vacant land and abandoned structures in seventy cities found that on average, 15 percent of a city's land was deemed vacant.[54] For a city with a rapidly growing population but fixed boundaries, vacant land and deserted buildings can represent a key competitive asset for economic development. They can create various kinds of jobs, increase tax revenue, improve infrastructure, and attract new residents, merchants, and money for improvement.[55] They can be reclaimed as opportunities for productive reuse, as solar farms, urban farms, community gardens, open land, general reclamation, and distribution centers for the future dwellers of metro centers.

Plans for reclaiming, stabilizing, and revitalizing neighborhoods will not only stimulate economic recovery and growth but will also help to eliminate a growing problem—arson. The US Fire Administration estimates that there are more than twenty-eight thousand fires annually in vacant residences and that 37 percent of these fires were intentionally set, resulting in $900 million in property damage and numerous deaths and injuries each year.[56]

The Department of Housing and Urban Development (HUD), Department of Transportation (DOT), Environmental Protection Agency (EPA), and several other agencies have made available hundreds of millions of dollars in funding to support the planning and implementation of projects to promote sustainable communities.[57] Federal and state grants for historic buildings have also helped finance these efforts. Investment funding is available for a variety of uses, including community planning, affordable housing, technical assistance, and capital infrastructure. To help navigate the complex maze of opportunities Reconnecting America has compiled a list of all upcoming programs and deadlines.[58]

WINNING THE WEATHER

In a smartly controlled building, comfort zones will be monitored by computers that will offer middle-of-the-road temperatures, or something like a constant humidity-controlled 72 to 78°F, depending on weather and, in living spaces, on one's age and gender.

A recent report by Christian Aid indicates that more than a billion people in coastal cities will be vulnerable to severe flooding and extreme weather due to climate change by 2070.[59]

The architecture group Terreform One adopts a counterintuitive but practical approach in its Governors Hook project in New York. Instead of keeping water out, the design allows the water in, to be stored or moved by many methods, from permeable pavement to redirection to building up. Many architects suggest preventing a siege mentality that would require people to fight a losing battle with the elements.[60]

A CHOICE OF HABITATS/A CHANGE OF HABITAT

The choices of where and how to live in the future will make the science fiction of yesterday morph into the facts of tomorrow. Imagine urban visions on the horizon of the future: living as a terrestrial on or under the earth, as an aquarian on or under water, as a citizen of the sky in massive skyscrapers, or as a space colonist in orbit or on other planets.

We can also do what ancients couldn't do with their cities—pick them up and move them. With developments in the assembling of buildings through drones, nanotechnology-enhanced materials, and industrial 3D printing, dissembling and deploying cities elsewhere could be accomplished with good planning. In fact, houses are currently being designed that can be moved by boat or dirigible.

Then again, one dystopian outcome is that cities will simply continue as they are or become deserted. The costs of change may result in some areas simply being sacrificed and abandoned. Unfortunately, the same may be true for people.

2

HABITATS FOR INHABITANTS

Home Is Where the Heart, Hearth, and Habitat Are

As cities grow and lifestyles change, the homes we decide to live in will change as well. In fact, we are already starting to see unique housing alternatives. Sometime in the very near future, we will see not only "smart" single homes but also superstructures that may encompass a dozen blocks or more. New forms and choices of housing, from the far-out to the hands-on, from movable homes to exotic homes in strange surroundings, will come in a plethora of shapes, sizes, and places.

Sixty years ago there were only two "megacities," urban centers with populations of more than ten million people: New York/Newark and Tokyo. Today there are thirty-three, although this number is expected to rise possibly to fifty-three by 2030, mostly in developing countries. In 1990 there were just ten megacities worldwide. At present there are about four hundred million people in megacities worldwide. And the future populations of megacities are projected to be

- 900 million in 2050
- about 1.1 billion in 2060
- about 2.1 billion in 2100[1]

All over the world, new metropolises are being built, offering a snapshot of what our future will be like. At present the US megacities are Chicago, Los Angeles, New York City, and the metropolitan area of Washington-Baltimore. It is anticipated that by 2050 Asia will have thirty megacities, and Atlanta, Miami, Phoenix, and Riverside-San Bernardino could pass the threshold by 2060.[2]

By 2050 some will be monster megacities of thirty to one hundred million. Figuring out where and how to house their denizens will take a humungous amount of work wrapped around heaps of options. A central theme is to plan to make homes cost- and energy-efficient, sustainable, and very comfortable—some might even say perhaps too comfortable, as housing accessories, robots, and apps may put an end to the toil of household chores.

Silicon Valley, the jewel in Santa Clara County, is a suburban enclave that has some of the most expensive real estate in the country as a result of tech companies making the area their corporate home. Unfortunately, the once sleepy community has become one of the toughest places to find affordable housing due to the influx of techies and well-to-do money.

Google and Facebook, two of the many companies that have inadvertently caused the housing shortage, are now—with the partnership of local politicians—doing something about it, perhaps as a precursor of the future, and they have led the way in a series of new proposals from big business seeking answers to the problems of cramped housing. More than one thousand potential additional units were approved by the city in which Google is located.[3] Hopefully this will be a hallmark of big business and community needs.

Google will invest $1 billion toward efforts to develop at least fifteen thousand new homes in the San Francisco Bay Area. "Across the region, one issue stands out as particularly urgent and complex: housing," CEO Sundar Pichai wrote in a blog post. "As Google grows throughout the Bay Area—whether it's in our hometown of Mountain View, in San Francisco, or in our future developments in San Jose and Sunnyvale—we've invested in developing housing that meets the needs of these communities. But there's more to do."[4]

THE FUTURE INSIDE

The new future home will be not only smart but also intuitive; all the devices and appliances will be connected by software and the internet of things (IoT), and interaction will come from a robot loaded with an artificial intelligence (AI), a Siri-like device that will take care of and anticipate most of people's needs.

As soon as you say something like "I'm out of laundry detergent" or "Do I have a clean shirt?" it will be picked up by a discretely placed microphone and trigger software to transcribe your words into a to-do list or into a command that will take appropriate action.

New research has found that 68 percent of people think their bathroom is old-fashioned and would like to see new innovations in this room of the home in the next ten years.[5] Some bathroom innovations on the horizon are included below.

Auto Body

A body analysis scale in a bathmat will not only track weight but also analyze your body fat percentage and body mass index while taking your temperature, blood pressure, heart rate, and thermometer readings from any number of trackers. These may be in the form of a wrist monitor, glove, body wrap, underwear, electronic tongue depressor, finger pinch monitor, among other utilities.[6]

All readings will be fed into a health app along with other data about your level of activity, nutrition, and sleep patterns to give a basic picture of your health, note any anomalies, and make basic health recommendations. This information can be relayed to a doctor or other health practitioner that can provide follow-up.[7]

Smart Mouth

Your smart toothbrush will connect sensors in the brush head to a health app, to give you real-time feedback on your brushing technique and tell you when to change your brush. Smart sensors will track everything from areas missed to whether you're brushing too hard; they will assess your gums and suggest changes that need to take place. They will also track tongue cleaning and oral health, and, since 80 percent of bad breath comes from odor-producing bacteria on the tongue, they will measure bacterial by-products in the mouth. The app then records the information, offers advice, and can send the results to your dentist.[8]

Toilet Training

Everything that goes into your body will be analyzed when it comes out, from its nutritional content to its caloric value. This will offer a complete picture of how well the body is functioning and how well you are living. An automatic urine or stool test will provide early warnings about urinary tract infections, pregnancy, and markers for types of cancer and other diseases.[9]

Vigilant for Viruses

Indoor air can be from ten to one hundred times worse than the air outdoors. Air purifiers equipped with professional grade sensors can monitor the levels of particles in the air. An air purifier can be set to check for allergens and filter out up to 99.9 percent of airborne viruses and bacteria that are 2,500 times smaller than the diameter of a human hair.[10]

Sampling Sleep Patterns

Obstructive sleep apnea (OSA) is a condition in which the walls of the throat relax and narrow during sleep, interrupting normal breathing and often resulting in snoring, broken sleep, and—at worst—asphyxia (a condition arising when the body is deprived of oxygen, causing unconsciousness or death; suffocation). Sensors will monitor sleep patterns and sounds and note abnormalities.[11]

Baby Watch

The latest technologies for babies' health will include connecting a smart baby monitor and ear thermometer to a mobile phone app that helps identify all manner of health matters, recording results for later examination. Babies' poop can be analyzed by automatic diaper or commode swabs. Results can be fed to a doctor's office. (This can also be used for adults.)[12]

Sensing Falls

If you've fallen and you can't get up, a help button pendant will sense it and activate a GPS locator and two-way communication with a helpline via a speakerphone that is automatically turned on and dialed by AI anywhere in the home. Movement sensors around the home will track all activity and behavior in general in the house. Smart analytics will identify each individual in the house and notice and record any abnormal behavior. Automatic phone or video calls can notify caregivers, physicians, or relatives. Of course, for those who are anxious about their personal privacy, the sensors and the personal information they store can be deactivated.[13]

Remembering Meds

Half of all medication for chronic illness is not taken as prescribed, especially by the elderly, often because of absentmindedness or dementia. So a smart medicine dispenser can issue reminders, via a watch, phone, TV, or in-house intercom device, to a patient and/or a caregiver when it's time to take meds. It can also calculate how many have been taken and share this information with the in-house medical app and a doctor, relative, or caregiver.[14]

Mirror, Mirror—and Apps

A look into the bathroom mirror will trigger a mini–health check and use facial recognition technology to pick up subtle cues about your health and mental state. It may advise you on what skin care regimen to use on a given day, based on your appearance.

Family Carebots

It is becoming increasingly important to devise ways of ensuring that the elderly, ill, and disabled can live safely and independently for as long as possible. One of their biggest issues is often lack of mobility— being unable to get out of bed, move around, take care of errands as well as themselves.

For such situations will be carebots, which are designed to provide assistance. They can range from life-size humanoid bots that can take blood samples and analyze them, lift patients, or help them walk; furniture that transforms from a bed to a wheelchair; and mobile servants that can fetch and carry things from one room to another and have the ability to monitor and notice the abnormal and communicate to various contact people.

There are hopes that robots will make aged-care jobs less demanding and help senior citizens maintain a longer independent life in their own home, assist caregivers at home or in a nursing home, or provide company to the lonely.[15]

Resourceful Refrigerators

Your kitchen app will take orders for beverages and foods from the pantry or refrigerator and can survey contents and make menus for days or weeks in advance. Miniaturized technology will allow the scanning of almost anything to find out its molecular composition and, in the case of fruits and veggies, check for ripeness. The smart fridge will be programmed to sense what kinds of products are being stored, keep track of what's been used, make comments about whether you're drinking too much beer or consuming too many carbs, order anything that is running low, and suggest nutritionally balanced meals; it will also sync perfectly with your levels of activity and weight-loss goals.[16]

Screen Tests

How about a TV that blends into the wall? That's the promise of a new generation of TVs. Samsung's Smart Things app, for example, allows you to take a picture of the wall using your phone camera, sends it to your TV, and reproduces your wall right on the screen. You can color-correct the image manually, but most of the process is automatic. Even when it is turned off, the TV displays the image of the wall. Also when off, the TV can still display news headlines, weather reports, and even traffic, or just show the time and date. Custom art is also included and can be matched to the background. [17]

Waving your arms at a gadget will turn electronic devices on and off, which might be "too smart" in an overly melodramatic or energetic household. Entertainment systems of the future will be able to plug into your moods—based on the number of hours that you spend watching TV or listening to music, how your choices affect your emotional vibe, and how you seem to want to feel when choosing a TV program, movie, or music. All can be tracked, remembered, and analyzed. Sensitive sensors will turn off a device if there is no one in the room for a period of time. [18]

Smarter Smoke Detectors

Smoke detectors and carbon monoxide monitors can alert you about dangerous levels of certain gases, but the newest technologies can let you know exactly where and when a problem exists and notify you about a fire, pollen levels, coming weather conditions, and indoor and outdoor air pollution so you can be sure to take appropriate actions. [19]

Let There Be Natural Light

Modern paints and materials will enhance natural light and reduce the energy needed for lighting in the home. Light for particular times of the day will be programmed and automatically regulated. Less energy used means lower greenhouse gas emissions, greater fossil fuel conservation, less waste produced, and a lowered utility bill. Smart window coverings and thermostats will complete the picture. [20]

A Biofuel Duo

Carbon dioxide (CO_2) emitted at home will be used to feed microalgae, producing biofuel that in turn will be used to generate heat and power. The CO_2 produced from the heat and power generation will be once again used for biofuel production in a closed loop. [21]

Nanobot Waste Watchers and Underground Collection

Liquid and solid waste material will be treated to break the matter down into its chemical properties for fuel or fertilizer. Consequently, it will be possible to recycle endlessly and re-create any type of material. Products for recycling will be sorted by microscopic nanobots that separate mixtures of materials into categories based on their size, shape, color and on their physical and chemical properties. This will be especially useful for colonies in outer space.[22]

Municipal waste will be collected via a pneumatic network separating and transferring the waste flow in underground tubes to treatment facilities in a 24/7 service, reducing the presence and pollution of vehicles that collect garbage in the city. Customer will be billed via an RFID (radio frequency identification) sensor by weight or material.[23] Multiple bins will be replaced by pods integrated underground. The pods will then drive waste via tubes to a treatment facility.[24]

THE FUTURE OUTSIDE

One example of really going urban green is a project in Singapore comprised of four towers, all connected by what's called the "heart center," an area filled with thousands of plants, trees, and even a waterfall. There are also sky bridges and terraces decked out in greenery, for a total of 160,000 plants helping reduce heat (by transpiration, reflecting sunlight, and creating shade) and improve air quality while reducing the amount of CO_2 on the island.[25]

The development is LEED-certified. ("LEED" stands for "leadership in energy and environmental design.") LEED certifies designs and structures that reduce CO_2 emissions and water and electricity consumption, and reinforce resource sustainability for buildings. It has four levels: certified, silver, gold, and platinum.[26]

Gardens

The garden will also experience a fusion of tradition and technology for plant lovers—multisensor gadgets will keep tabs on everything from water content and soil acidity to temperature, fertilizer, and ripeness, in the case of fruits or veggies. Meanwhile, robot mowers and pruners will keep the landscape neatly trimmed, and digital art will allow the stylization of the garden with beautiful and changeable holographic statues, colors, and sound scopes.[27]

Microchips and remote controls are becoming as popular in the garden as they are in the kitchen, den, or even the office. We knew it was only a matter of time and that time is now. Smart-thinking landscape architects are reducing their environmental impact via water conservation, solar technology, kinetic energy, and drought-tolerant planting—at the same time creating charming designs.[28]

Accommodating Accommodations

New technologies in building and materials will allow construction of previously unheard-of geometric complexity. Design trends will become more free-form and organic. New shapes and new combinations of materials will allow for entirely new building aesthetics and larger structures.

Finally, cost will be cut by robots handling heavy and dangerous work, inexpensive prefab building components, and the use of existing material (like dirt and clean refuse), all of which will afford architects more creative leeway.

Prefab components are also environmentally friendly, as they reduce material waste as well as the number of delivery trips needed to the construction site. In other words, instead of transporting raw materials and basic supplies to the construction site to build a structure from scratch, most of the structure is prebuilt in a centralized factory and then shipped to the construction site to simply be assembled.[29]

Along with techno-changes, adjustments will have to be made in long-standing and out-of-date zoning and building ordinances that have often impeded development and innovation. If urban areas are going to attract developers, then city planners must recognize the need to overhaul outdated policies concerning land use and built environments. Cities need to consider a range of innovative and aggressive polices to lure new money and make room for more people.

The dream of universal affordable housing has been an idea tried and tested by architects throughout history, from Bucky Fuller's wacky Dymaxion House,[30] to mail-order homes assembled like do-it-yourself furniture. This isn't to say that poverty, ghettos, and disadvantages won't be part of the landscape, but we will have the technology and, hopefully, the will to begin removing those roadblocks.

Some of the promises of 3D printing—one example is "Contour Crafting Technology"—is the capacity for building multiple homes quickly, creating less waste than conventional construction methods, and the use of robotics for labor. Of course, that's bad news for the people in the construction industry who will suffer the loss of thousands of jobs.[31]

With any luck, owning a home will no longer command the sizable investments of generations past when younger buyers were priced out of the housing market. On the other hand, a glut of new housing will begin lowering housing prices, negatively impacting current homeowners who are depending on the stable or rising equity of their homes for retirement or the ability to move on up.[32]

Chic Choices

When it comes to the price of homes, it shouldn't come as a surprise that the majority of the sticker shock comes from the value of the land more than that of the actual structure. An even bigger factor driving the value of land is the demand for housing within a chosen location, which could cause the housing market to boil over.

In the future, a lot of folks will be buying movable micro homes and even small apartments that can be parked, like cars, on small lots. Some even imagine that people will be able to move their houses from city to city with the aid of autonomous and remote-controlled aerodrones. (A remote-controlled vehicle is always controlled by a human. Autonomous devices are aware of their environment and incoming data and have the ability to learn and make decisions on their own. By 2020, an estimated fifty billion of these devices will be connected to the internet.)

Some homes in the future will downsize in terms of square feet, but there will be many more choices available. One example is Ecocapsule, a smart, self-sustainable micro egg-shaped structure that utilizes solar and wind energy for power and a battery for storing energy. Rainwater is collected on the surface and filtered into a water tank. It enables the inhabitant to stay in remote places in comfort and can serve as a cottage, pop-up motel, mobile office, or research station. It's been engineered to be self-sufficient, practical, and functional. It measures 15.32 feet by 22 feet, is 8.20 high, and weighs about 35,000 pounds with a full tank of water (and filtration system) and incineration toilet. It is made of fiberglass over a steel frame.[33]

Homepod specializes in building energy-efficient homes that the company expects will be popular among new homeowners looking for more efficient, smarter homes with controls for electricity, security, HVAC, and more. The houses range in size from two-bedroom apartments to four-bedroom townhouses.[34]

Tiny houses are becoming a popular alternative as well as a new social movement. People are choosing to downsize, live with less, simplify the space they occupy, and still be comfortable. Tiny houses can be configured for a shared, two-person household or for an extended family. Most range from one hundred to four hundred square feet; among their features are movable partitions, recessed-in-the-wall amenities, and pull-down furniture, which enable transformation of the unit without reconstruction. Countertop and cabinet height may be adjusted manually or mechanically for all customer sizes and physical abilities. There are many no-cost blueprints available online. More and more cities are rethinking lot size regulations to accommodate tiny homes and communities. [35]

One of the most creative models is an experimental, low-cost, micro housing unit made from concrete water pipes approximately eight to twelve feet in diameter, and eight feet wide. The repurposed pipes are designed to accommodate one or two people and come with approximately one hundred square feet of petite living space. The interiors are made up of micro living room furniture with a built-in bed, a mini-fridge, bathroom, shower, and plenty of storage space for clothes and personal items. [36]

Although these structures are not lightweight, at twenty-two tons they require little in terms of installation costs and are easily stacked. Entire tube communities could be installed in small, unused spaces. Cost varies around $15,000 (not including the cost or use of land). [37]

In its approach, SPACE10 shares goals and methods similar to Wiki House (a generic name for an open-source project for designing and building houses that endeavor to democratize and simplify the construction of sustainable, resource-light dwellings). SPACE10 focuses on something that, while not terribly new, has rarely been explored. Known for simple, well-designed, flat-pack furniture, IKEA is proposing expanding their DIY model to a much larger scale: entire city centers with square boxes shaped and stacked on one another. [38] The result is low-cost, adaptable, and sustainable housing that could be manufactured locally. It's a 527-square-foot micro house built using only a milling machine and plywood certified by the FSC (Forest Stewardship Council), with a total cost of around $206 per foot. [39]

The ALPOD is a sleek, rectangular 42.6-by-10.8-foot mobile home (about 480 square feet of living space) made of insulating blocks and wooden panels with an aluminum sheath, with thermal insulation, solar power, and a built-in kitchen and bathroom. [40]

The Pop-Up House is low-cost, recyclable, passively heated, and has all of the qualities of tomorrow's homes. The prefab structure snaps together like LEGO bricks in a few days. Robotic labor may increase the speed of this construction, improve its quality, and lower its price. [41]

A "digital construction platform" is another mode of 3D autonomous printing that boosts efficiency and building strength; it only puts material down where it's needed, and it's safer, faster, and more precise than manual construction methods. [42]

We will also see more and more apartments in cities that have converted parking garages and repurposed other buildings to living spaces comparable to tiny houses. Some of these retrofitted garages have a community feel with amenities that urban folks find attractive and venues for theater, dining, and other cultural activities desired by those who want to live, work, and play in the heart of the city.

Hands-On DIY

Do-it-yourself (DIY) projects (like the Wiki House) are being developed by architects, designers, engineers, inventors, manufacturers, and builders, collaborating to develop the best, simplest, most sustainable, highest-performance building technologies that anyone can use and even improve upon. [43]

The developer's aim is to offer designs to every citizen and business and to make it easier for industries to deal in, invest in, manufacture, and assemble better, more affordable homes in order to grow a new housing industry while reducing dependence on the top-down, debt-heavy mass housing systems of the past.

Fab Prefab

It took only three weeks for a Chinese company to build a fifty-seven-story skyscraper. The construction company reportedly wants to try to build a 2,749-foot skyscraper. The "Mini Sky City" tower is the work of Broad Sustainable Building (BSB), a Chinese firm that specializes in prefabricated construction.

By preparing more than 2,700 modules in a factory for four months before site work began, BSB says it was able to assemble the structure at the rate of three stories per day—like a giant vertical jigsaw pieced together from a minutely detailed set of instructions. [44]

The company already boasts a fifteen-story hotel assembled in six days and a thirty-story hotel in fifteen days, among other achievements. What the Chinese are trying to do is sell buildings worldwide—making them in China and shipping them across the globe. [45]

One of the obvious pros of using modular construction to build affordable prefab housing is that units can be assembled off-site and then quickly stacked into place. It's another alternative that significantly reduces development time and cost, making it easier to build affordable housing faster and cheaper along with smaller microapartments and multifamily buildings.

There are many prefab houses on the market that are made of various materials from fiberglass to metal and foam over a frame. Even HUD is looking into financing prefab homes for the growing market. [46]

Biomimicry in Buildings

It's one of the targets of futurist buildings—the use of geometry to assemble and repair structures that will grow and evolve all on their own, like trees, assembling their matter through something like genomic instructions encoded in the material itself. It's a contemporary philosophy of architecture that seeks solutions for sustainability in nature, not by replicating the natural forms but by understanding the rules governing those forms.[47]

A major problem worldwide is power shortage paired with buildings' high consumption of energy. As they attempt to resolve this issue, architects are turning to biomimicry, which simulates or co-opts processes that occur in nature, producing, for example, ultrastrong synthetic spider silks, adhesives modeled after gecko feet, and wind-turbine blades that mimic whale fins.

Biomimicry figures in the belief that architecture should reflect the geography and culture of its setting and that architects must discover the most efficient solutions that resemble available natural objects. They might build screen systems on windows that use elasticity, geometry, and thermobimetal properties to open and close in response to sunlight—as flowers do. Inspired by coral reefs, researchers are using bacteria that alter the pH balance of surrounding material in order to allow calcium carbonate to grow and bind the material together with little outside energy and no carbon emissions.[48]

Unused Land and Vacant Buildings

On average, 15 percent of a city's land is considered vacant; it ranges from undisturbed open space to abandoned, contaminated structures to brownfields.[49] The roots of today's hypervacancy problem lie in the Great Recession of 2007 and subsequent foreclosure crisis, especially in inner cities.[50]

One of the worst examples of hypervacancy is the 84,641 blighted structures and vacant lots in Detroit, almost half of which should be demolished, which would cost almost $2 billion.[51] These Detroit properties collectively form a space the size of Manhattan, and they're not alone. Gary, Indiana, has 25,000 vacant homes or lots, covering 40 percent of the city's parcels, and Philadelphia found 40,000 vacant lots with no known use. Vacant structures in the country number more than twelve million.[52]

Many cities contend that when a structure is abandoned, it presents an "imminent danger" to the community and threatens the city's "health and safety," attracting arson, squatting, drug use, and other illegal activities.[53]

Abandonment or vacancy is a sore spot for many communities regardless of size and geographic location. But it also represents a lot of possibilities for use, from inner-city farming to storage to repurposing structures for public housing or community activities—or for soup kitchens for the homeless and hungry, as famed Italian chef Massimo Bottura has done.[54]

Some cities are tearing down and/or transforming homes to create affordable single-family neighborhoods. Other cities with local nonprofits have turned to greening these buildings and lots, creating urban farms, pocket parks, playgrounds, and community gardens.

Design advocates are encouraging redesigning and retrofitting existing buildings rather than building new. Major renovations and retrofits reduce operation costs and environmental impacts and can increase resilience in a neighborhood. An existing building should be looked at in terms of the human labor and material costs that might be saved by renovating it rather than demolishing it and constructing a new building. Retrofitting an existing building can often be more cost-effective than building a new facility.[55]

People want buildings that inspire and delight. Inside buildings and out, designs should try to realize stunning effects and playful forms in order that buildings work in harmony with their surroundings and their residents appreciate their living spaces.

URB PRICE OUT

A good example of urban price-out is going bicoastal. In San Francisco following the collapse of the housing market in 2007, the median price of a home was around $700,000 compared to a whopping $1.65 million.[56] Manhattan has witnessed a 20 percent raise in rent just since 2016.[57]

Many cities (like New York and San Francisco) have no more space to build. Middle-class employees like teachers and office and city workers have been priced out of the city. And they can't look to the suburbs for relief—the burbs are bursting at their borders with boarders looking for a break, and the cost of either buying or renting is unbelievable and unrelenting.[58]

OLDER HOMEBODIES IN THE FUTURE

As a generation the millennials are hesitant about making babies and buying homes. They are not in a financial position to dive into the current housing market with average listings costing a millionaire's ransom. The nation's birthrate is going down and at present, there are more baby boomers and seniors in the suburbs than there are millennials with children, leading in some places to the decay of tax bases.[59] The aging boomers care less about schools and parks than they do about supporting services that they'll need as they age. The traditional suburbs are in danger of becoming senior burgs.[60]

In response cities are starting to go "country," providing a lifestyle for people who value walkability, sustainability, green surroundings, and the ability to live within their means. Urban city planners are starting to redesign the landscape so that people can ride their bikes and enjoy nature and space. The property lines of city and country areas are becoming blurred as modern population's values and bank accounts become modified or completely transformed. Everywhere in these new communities, residents prefer driving a mile or two instead of ten or twenty, and they own one car instead of two.

"Access, access, access" to all the things that make cities great places to live has become at least as important as "location, location, location."[61]

Metro markets have to deal with changing demands not only for living and working spaces but also for resources—water, energy, air, food, transportation, city services, sustainable building methods and materials, green spaces.

GREEN CERTIFICATION SYSTEMS

In the future LEED may see heightened competition in new construction ratings from the Green Globes (GG) rating system and possibly from new entrants in specialized niches, such as retail or office interiors. In 2013 and 2014, the federal government put LEED and GG on an equal footing for government projects, lending further legitimacy to GG.[62]

While both LEED (the most widely used green building rating system), and GG (which promotes a sustainable future and a healthy planet by designing homes with green energy solutions) seek basically the same goals and ideals, there are differences. LEED calls for a minimum indoor air-quality performance while GG does not. LEED makes it

mandatory that builders have "some documentation of the initial building energy and operational performance through fundamental commis-
sioning." In the United States LEED is run by the Green Building
Initiative (GBI), a nonprofit organization. The LEED rating system
frequently requires prerequisites to many of their credits, whereas GG
does not require any prerequisites. It should also be noted that GG uses
life-cycle assessment and multiple attribute evaluations, whereas
LEED does not use them.

The LEED process is also far more stringent than GG in a few
areas, but GG is a lot more user-friendly. LEED has minimum standards that must be met in order to begin the certification process and
requires detailed documentation for every point pursued.[63]

GG is used to certify a wide variety of building types, including many
that cannot be certified through LEED. Examples include recreational
centers, transit centers, and parking garages, to name a few. It is increasingly becoming the system of choice for building owners, managers, architects, and engineers who want an alternative that offers the
quickest and most understandable way to achieve superior building
performance.[64]

The Building Research Establishment Environmental Assessment
Method (BREEAM) is the world's leading sustainability-assessment
method for master planning projects. Its marketing system is used in
sixty countries and seems poised to enter the United States.[65]

Other North American systems include the Living Building Challenge and LEED Canada, which competes in existing buildings with
the Building Owners and Managers Association (BOMA), which ad-
dresses an industry need for realistic standards for energy and the environmental performance of existing buildings based on accurate, independently verified information.[66]

GREEN DESIGN: TOO MUCH GREEN?

In a survey of more than seven hundred construction professionals, 80
percent cited "higher first costs" as the biggest obstacle to green building; it's the most common criticism of sustainable building,[67] despite
claims that "LEED buildings cost 25 percent less to operate and enjoy
nearly 30 percent higher occupant satisfaction and lower interest
rates."[68] And overall, "the more green building and materials are used,
the more the cost will lower."[69] A 2003 study by the California Sustainable Building Task Force shows that an initial green design investment
of just 2 percent will produce savings greater than ten times the initial

investment, based on a very conservative twenty-year building life-span.[70]

A study of twenty-two green federal buildings was conducted by the General Services Administration and the Pacific Northwest National Laboratory. It compared one year of operating data and surveys of green building occupants to those of the national average for conventional commercial buildings. It found that green government buildings

- cost 19 percent less to maintain
- used 25 percent less energy and water
- emitted 36 percent less CO_2
- had a 27 percent higher rate of occupant satisfaction[71]

Others maintain that the initial costs simply outweigh benefits. The costs associated with these structures are believed to be quite expensive. In fact, homeowners might have to invest lots of money; however, in the long run, the invested money could be returned through energy-saving possibilities.[72]

Even if green construction does cost more as an additional investment, it typically yields operational savings worth several times that much. Other studies show that many LEED-designed buildings do not cost more and can actually cost less than conventional construction as they save money in the long run via their sustainable practices.[73] Numerous sources of funding for green building are available at the national, state, and local levels for industry, government organizations, and nonprofits. To begin, check out https://archive.epa.gov/green building/web/html/funding.html.

A CHANGE OF HABITAT

We will be able to do something the founders of the first cities could not—pick up and move them. With developments in the assembling of buildings through drones, nanotechnology-enhanced materials, and industrial 3D printing, dissembling structures en masse and deploying them elsewhere could be accomplished with good planning. At present, houses are being designed that can be moved by boat, dirigible, or drone.[74]

Then again, one dystopian outcome is that cities will simply continue as they are or be deserted. The costs of change may result in some areas simply being sacrificed and abandoned.

In a smartly controlled building, comfort zones will be monitored by computers that will offer middle-of-the-road temperatures, or something like a constant, humidity-controlled seventy-two to seventy-eight degrees depending on weather and, in living spaces, also depending on your age and gender.[75]

ZERO NET ENERGY: A PLUS

Zero net energy (ZNE) consumption means that the total amount of energy used by the buildings on an annual basis is roughly equal to the amount of renewable energy used that produces no greenhouse gas. ZNE buildings should be the goal of all cities, present as well as future. Developers of speculative commercial buildings have begun to showcase ZNE designs.[76]

As we venture out into the solar system, many new modes of living space will be experimented with and improved upon, incorporating sustainability, comfort, and planned growth. Space is the limit.

3

BURROWING BENEATH THE EARTH

Tomorrow's Troglodytes

There are cities beneath the streets.

—Robert E. Sullivan Jr.

The prospect of living underground for long periods was the quirky plot of a 1999 movie comedy called *Blast from the Past*, about a family that hid out in their bomb shelter in the 1960s, naively mistaking a plane crash on the surface for a nuclear attack.[1]

Another was the nightmarish *The Time Machine*, by H. G. Wells, where the earth's people divided themselves into surface people and those staying underground, with horrific consequences.[2] Going underground—literally—has been the topic of much fact and fancy and has gained a lot of traction of late.

In a few short decades the world will face epic challenges that might not be as quirky as dystopian fiction, but the population bomb, rapid urbanization, potential for nuclear conflict, and the demise of our biosphere are just some of the challenges impacting our metro areas. When looking at the cities of the future, we have to look for more than just quick fixes; we need bold new ways based on radical ideas, and planning methods in which to pursue them.

With close to 90 percent of the world's population projected to live in cities by the end of the century, maybe the only way for future cities to go is . . . down.[3] The idea that roads and support services must be on the ground level of a building is antiquated and ridiculous. A 2013 report by the US National Research Council suggested that "underground facilities may be the most successful way to encourage or support the redirection of urban development into sustainable patterns."[4]

GROUNDBREAKING UNDERGROUND

What goes up can also go down. Some predictions are that two-thirds of the world's population will be beneath the ground—and alive—by 2050. From subterranean parks and malls to inverted skyscrapers and tunnel-farms, these hidden underground cities and urban projects are going to have us spelunking in some form in the future.[5]

The urban underground is full of deserted shelters and bunkers—remnants of past cold, warm, or hot wars—air raid shelters, bomb silos, subways, storage depots, catacombs, caves, and natural caverns. Most of these single systems and complexes are now empty, but they still exist and they can be utilized, especially in emergencies. Maintenance in their present state costs a lot of money, but there's a lot of potential to turn a problem into an opportunity. There is plenty of room down there for living spaces, as the deepest mines are much deeper than the tallest buildings are tall.[6]

The success of "building down" might depend on helping people overcome fears associated with small, dark, claustrophobic environments, with being buried alive and not having a clear way out, or with flooding and fires.[7] But there are ways to counter such fears, like connecting all areas of an "earthscraper" to a large, central, seemingly open space that receives light and air from above as if it were a canyon, with clear oases, palm trees, and perhaps illusions of the sky (see below).[8]

Let there be no mistake: living underground is a huge environmental and economic issue. But underground spaces are less susceptible to external influences like natural disasters, and their impact on the internal environment is less than on aboveground facilities.

Lack of Light

Dr. Lawrence Palinkas, chair of the Department of Children, Youth and Families and professor in the departments of social policy and health, anthropology, and preventive medicine at the University of Southern California, says a lack of sunlight can cause difficulty with sleep, mood, and hormone function, which can produce chronic diseases of several varieties. But, he says, "timing and routine exposure to bright light that can mimic the properties of sunlight might enable people to live underground for long periods of time."[9]

Nonviable indoor artificial lighting, like casino lighting, is inappropriate because it causes people to lose their sense of time; and no difference between day and night is evident. When the sun goes down, the brain sends signals to our body to secrete melatonin, which is the sleeping hormone, and when the sun rises, cortisol is released to make us get out of bed.[10] On the brighter side, artificial lighting allows businesses to operate twenty-four hours a day, because workers can come and go at any time without worrying about being confused by sunlight or moonlight.

The Light Green Scene Sham

The "fake nature" of underground ecology in underground plazas featuring bogus palm trees, phony tropical flowers and ferns, and suspicious moss are nothing more than cheap exoticisms that eventually gather dust, decline, and wither, looking rather sad. The scene doesn't fool anyone, and—what's worse—it is so artificial that it reinforces the fact that you are underground and away from real nature.

Consequently, sustainable underground cities must have environments designed to look real—or genuine facsimiles—of subterranean life. Daylight will be provided by electrical light sources and in some places by natural skylights; these can focus light and reflect or mirror light to certain areas for growing things and furnishing natural sunlight.[11]

Subsurface Streams

Instead of holding onto the idea that water threatens to flood, causing mold, undermining foundations, and attracting vermin, make water a friend instead. Freshwater may very well become a threat when natural systems are mistakenly tapped.[12]

Clean freshwater, while circulating, can enhance the landscape, creating waterways that become small rivers and creeks. These in turn can be oriented to flow in ponds and pools, aiding plant and animal life and creating a more natural setting and atmosphere. Some of the largest freshwater accumulations are in underground aquifers.

"EARTH-SHELTERED" LIVING

While under-the-earth housing is not common today, it has been around for a long while. Ancient caves were our first natural shelters.

From the 1964 World's Fair

"Greater security, peace of mind, and the ultimate in privacy" were the words used to describe a revolutionary home fifteen feet below the surface that was showcased at the 1964 expo in Flushing, New York. It was devised as a prototype for future residential design where the homeowner could create "his own private world," shut out the dangers of aboveground living—intruders and storms, for example—and control the home's environment by dialing in any climate electrically.[13] Jerry Henderson, the founder of Avon Cosmetics, originally funded the model; later, builder brothers Jay and Kenneth Swayze provided funds as well.[14]

The builders had become interested in subterranean living when they were contracted to build a bomb shelter in Plainview, Texas. Afterward, Underground World Homes were featured at the 1964 World's Fair, and they garnered a lot of interest. However, the brothers came to realize that the cost of building a home underground was prohibitive for the masses and there wasn't enough market for the proposal. In 1978 entrepreneur Jerry Henderson and his wife, Mary, continued Underground World Homes. The Las Vegas model home measured nearly fifteen thousand square feet, and the house sat twenty-six feet below the surface. The couple also started an underground-building company called Geobuilding Systems Inc., but they shut it down in 1980.[15]

✻ ✻ ✻

During the energy crisis of the 1970s, people became increasingly interested in finding ways to reduce their consumption of energy and fuel with energy-efficient forms of architectural design. Others thought to go underground to escape hostile weather conditions. Subterranean living seemed to offer great promise, and experiments were undertaken to explore potentially practical options for homes below ground that would be affordable and energy efficient.[16]

Although the efforts met with mixed success and the underground movement lost momentum, it never completely died out. As we have seen, a lot was learned about what works and what doesn't when going down to live in the ground. As a result, there are a lot more workable options from which to choose today.

PROS AND CONS OF BEING BELOW

As in many instances, choosing where and how to live offers a myriad of alternatives and answers. One person's pros may be another's cons. What this book desires to do is offer reasonable and workable choices— which ultimately are yours.

Pros

- Less company: people will be less inclined to invade your front yard or sit on your front porch without an invitation.
- Protection from natural disasters except earthquakes and flooding.
- Doubly effective insulation: underground homes heat up nicely during the day with average sunlight, and hold heat extremely well overnight. And they can be cool in summer. Because the surrounding temperatures are so mild, heating and cooling costs can be reduced from 50 to 70 percent in an earth-sheltered residence. However, this will require the use of a lot of insulation; unprotected walls will eventually reach thermal equilibrium with the surrounding earth unless steps are taken to ensure that the heat produced or collected inside the home is not leached away through the walls.[17]
- Minimal cost to heat and almost nothing to cool: starting from an average base temperature of a cool 55°F, underground homes easily reach and hold a warm temperature of 75°. Those that are heated solely with a renewable resource will cost only about $500 a year to heat.
- Consistent temperatures, due to the properties of the ground's thermal mass: the planet's natural warmth can be exploited as a source of geothermal energy. The ground beneath our feet has a higher thermal mass than just about any other substance. The earth's capacity to store heat for a long period of time is quite impressive.[18]
- Lighting choices: many people are surprised that an underground house can be lit as extensively as conventional houses with lots of natural light.
- Blocking of aboveground street noise.
- Few chores: cleaning gutters and mowing lawns are not required.
- Generally cheaper insurance due to the level of shelter.[19]

Cons

- Buildings can leak and flood easily causing mold, water damage, and, at the extreme, evacuation.
- Limited landscaping: it's tough to grow plants underground.
- Possible cracking and crumbling during earthquakes.
- Difficult and expensive repairs: many ordinary materials don't hold up underground.
- Pest problems as bad as or worse than in an aboveground home, as a lot of potential pests burrow or live underground.
- High heating and cooling bills during extreme temperatures: these can run as much as 30 percent more than a conventional, well-insulated house of comparable size.
- Some differences in depreciation compared to aboveground conventional homes.
- Extra lighting needs: some lower floors would need lighting supplied by fiber optics or some other sort of energy source.[20]

SPECIAL UNDERGROUND ASSEMBLY ISSUES

Thin sheets of waterproof material specially designed for home protection will need to be applied to the walls and the roof to make an earth-sheltered residence watertight, since the earth can easily pass on moisture to anything with which it comes in contact. A special type of drainage or filtration mat or system will need to be placed over the insulation on the roof to ensure that any moisture that comes from above can be easily channeled away.[21]

Underneath the poured concrete foundation of an earth-sheltered home, a layer of sand at least four inches deep (for the purposes of drainage) should be put in place so that water cannot work its way into the home from below.[22] A living roof made of soil and vegetation can provide even more protection from the elements that will be absorbed by the roots of the green roof's plant covering before it can seep in deeper and cause trouble.[23] And it can make for an additional garden.

To keep the insulation from touching the earth, a protective layer of waterproof wood, hardy board, or plastic will have to be added, and the layer used must be thick, strong, and well coated with preservatives to withstand moisture and the earth's pressure without warping or breaking.[24]

Water Watch

The correct choice of location for earth-sheltered structures will go a long way toward eliminating any potential water-related problems. Most importantly, it is essential to always build above the water table; otherwise even the best waterproofing schemes will be tasked to the breaking point, and it may prove all but impossible to keep moisture from leaking into and through buildings.

The weight and pressure of the earth will obviously put an enormous strain on the walls and roof of a below-the-earth edifice, which is why concrete makes an excellent choice for a building material. While 3D concrete printing can work just fine for the foundation and the floor, concrete blocks stacked and coated with a thin layer of fiber-reinforced surface-bonding cement may be the proper selection for its walls.[25] Heavy timbers should be used to construct the roof, which must be able to handle the combined load of the earth and any other built structure above or on it.

Even though underground living can be highly efficient, it is unconventional and has some special requirements. The costs of a home constructed in this style may generally run from 10 to 30 percent higher than the average aboveground structure.[26] Ultimately, the return on investment provided by lower fuel costs will help to negate any extra up-front costs. Studies have shown that over the long haul, earth-sheltered buildings are the most economical for those living in climates that have extreme temperature changes and low humidity. Homes tend to pay for themselves quickly in those locations where relatively long, cold winters and scorching summers are the norm.[27]

Radon Raid

One hidden problem that potential underground homeowners must be aware of is the possible presence of radon, a colorless and odorless gas, produced in the ground by uranium decay that can be life-threatening if it collects in sufficient concentrations. When you have a home that is located underground you are much more likely to get deadly seepage in through the walls and floors of your home. Your home can act like a catalyst for the radon to collect and build up over time becoming a deadly component of your indoor air.[28] While it is not impossible to build a safe earth-sheltered structure in areas with elevated radon levels, steps will have to be taken to guarantee that all radon can be collected and vented from the home, which adds another layer of expense onto a project. Inexpensive radon warning meters are available in home-improvement stores.[29]

Earth Berming

While some earth-sheltered homes are built completely beneath ground level, many are constructed using a technique known as berming, where piles of soil are pushed up against the walls to form a protective cocoon of earth and vegetation that will separate the outer shell of the home from the elements. Roof covering of soil and vegetation offer more protection against atmospheric heat and cold.[30]

There are three primary design styles for earth-sheltered homes: atrium, penetrational, and elevational. For those who would like to maximize their protection from the sun, wind, heat, and cold, the covered atrium style (a large enclosed central skylight surrounded by a building) is definitely the way to go. In this type of arrangement, each of the rooms of the home will face the atrium from the north, south, east, or west, with spacious windows and possibly glass doors to allow the natural light to filter in from above.

Elevational houses are built by burrowing directly into hillsides or mountainsides and look almost as if they have been inserted into the earth; their side and back walls are completely covered. The front of such a house is left open to the air, usually facing the south in order to harvest the natural heat and light provided daily by the sun. Elevational homes are the least expensive type of structure to construct, and with their hillside locations, they frequently offer grand panoramic views.

Penetrational houses are built aboveground but are designed to fully exploit the protective abilities of the earth. Each wall of such a house is completely bermed with only the spaces over doors and windows left open, to facilitate good cross-ventilation and the effective harvesting of natural light. Variations on the penetrational approach are certainly possible; for example, the southern side could be left open as in an elevational home, while the rest of the house (save for the windows and the back door, of course) would be fully bermed.[31]

STOCKPILING SUPPLIES

Water

There should be plenty of water underground, but drilling for artesian wells must be carefully done as it could cause flooding. Groundwater in aquifers between layers of poorly permeable rock, such as clay or shale, may be confined under pressure. If such a confined aquifer is tapped by a well, water will rise above the top of the aquifer and may even flow from the well onto the land surface. It's like water in a plastic baggie. When you push a straw through the opening and then squeeze the baggie, water will gush out through the straw.

Food

Lots of things can be grown underground (think "underground" marijuana merchants), especially with current methods of deflecting sunlight or using solar-powered grow lights. Through hydroponics, growing food would not be a problem; aquaponics could yield fish and fowl; and eggs could be easily available. (See chapter 10.)

Energy

Solar energy could easily be gathered on the surface to generate and relay power below, or it could be directed through mirrors on the surface to do the same via photovoltaics or via a steam-operated turbine.

Wind farms could also generate energy from above and either direct it for use below or store it under the surface. Geothermal energy would be a no-brainer once a city is established beneath the earth with the capacity to generate power from the earth's heat. (See chapter 9.)

ARCHETYPES OF UNDERGROUND LIFE

Actually, the United States has a rather deep and checkered history of being underground. In New York and Chicago, there was a sub-rosa, subterranean society of bootlegging and speakeasies run by the mob and crooked or loose politicians who wanted their dalliances to remain private.[32] Also underground in Chicago was the place that Enrico Fermi and associates first brought about a nuclear reaction, which led to the invention of the first thermonuclear weapon.[33]

The Low Lowline

Located in one of the least green areas of New York City, this project is aimed at transforming a vacant trolley terminal into an underground park well below street level. The project is currently in the works and is expected to open in 2020. People will stroll at their leisure and appreciate the splendor of underground nature while a periscope-like system channels the sun's rays downward through the roof. Thanks to this "remote skylight technology," the park won't feel like a troglodyte re-treat.[34]

South of the Border Underground

In Mexico City designers have created a plan for the "Earthscraper," a seventy-five-story subterranean city intended to bypass Mexico's stringent building regulations and the city's growing space problems. The inverted pyramid will descend nearly 1,000 feet (304 meters) below the surface and potentially house up to one hundred thousand people, with terraced floors receiving natural light from a huge glass ceiling above. The lower floors will need extra lighting furnished by fiber optics.

The enormous complex will house a museum and ten floors of affordable housing. The rest will be made up of commercial office and retail space, at an estimated cost of $800 million. But because the idea is so new, there are no laws or guidelines for building downward in a city area where such activity could jeopardize the surrounding historic buildings.[35]

The RESO, Canada

The largest subterranean complex in the world, the RESO has been in use since the 1960s and is visited by more than five hundred thousand people per day, especially to escape Montreal's harsh winters. Visitors marvel at its twenty miles of tunnels spread over an area of seven and a half miles of downtown Montreal. It contains hotels, restaurants, galleries, stores, rail stations, cinemas, nightclubs, sixty residential complexes, a library, and a hockey stadium.[36]

Underneath Helsinki

The Finnish capital's "Underground Helsinki" includes more than 125 miles of tunnels as well as plans to continue to expand subterranean public spaces. Thanks to the many buildings' rough, rock-carved walls, the complex is also a pretty awesome architectural landmark. The city has plans for two hundred more under-the-earth structures.[37]

SubTropolis City

An eccentric underground project in Kansas City occupies 1,100 acres of abandoned limestone mines. More than 1,600 people work in this naturally climate-controlled, rock-carved space. SubTropolis houses everything from retailers, manufacturing firms, consumer products companies, auto storage, and an array of small businesses. "Six million square feet of it is ready, and we have room to build out another eight million square feet based on demand," said Dick Ringer, SubTropolis's general manager. [38]

Subterranean Singapore

With a population of nearly 5.5 million people squeezed into a small city-state, the main thrust for going underground is to help solve the land shortage issue while preparing for a growing population. Below the surface of the city is a 3,229,173-square-foot research and development facility that will support biomedical and biochemistry industries, among others, along with restaurants and shopping malls and below-ground residential living areas. [39]

Cool Down Under

In Coober Pedy, Australia, temperatures can reach 127°F aboveground, but below the surface the temperature is around a mellow 72°. Coober Pedy bills itself as the "opal mining capital of the world," but the town holds another more peculiar distinction: it may have more cave dwellers per capita than any other place on earth. [40]

Moose Jaw, Canada

Beneath the streets of Saskatchewan lurks an extensive network of tunnels that had been adopted by the Chinese as living quarters and workplaces because they were cheap to run and the weather there was not hostile (unlike the 1920's Canadian Ku Klux Klan). Rumor has it that Al Capone also used the tunnels for smuggling. Now there are theaters, shops, restaurants, and museums in the space. [41]

Cheyenne Mountain

Once home to the North American Aerospace Defense Command (NORAD), this underground city was built beneath Colorado's Cheyenne Mountain to hold thousands of people in the event of a nuclear attack. Terra Vivos, builders of survival communities, is working to create a series of underground complexes made to withstand anything from a 20-megaton nuclear blast or 1,250°F fire to 450-mph winds or a magnitude-10 earthquake. The shelters are outfitted with enough food, water, clothing, fuel, and medicine—and a wine vault—for all inhabitants to survive a year.[42]

Below Beijing

Dìxià Chéng, the underground city, covered an area of more than twenty thousand acres and was created as a city-sized refuge from nuclear attack. Approximately three hundred thousand people worked by hand and carried out everything they dug in bamboo baskets. When work began in 1969, the government boasted that the city could contain around six million people. The various tunnel systems linked up around ten thousand atomic bunkers that eventually housed restaurants, theaters, warehouses, factories, a mushroom farm, and sports facilities. Most of Beijing's underground world was privately owned.

In 2010, Beijing municipal authorities announced that the residential use of underground spaces would be illegal by the end of 2012, citing safety hazards, such as the risk of fire or flooding. In 2015, thousands of residents were evicted from their underground housing.[43]

Under Iran

The city of Kariz-e Kish is more than 2,500 years old and was initially referred to as the system of aqueducts of Kish Island. Today, it is transformed into an amazing underground city, with an area of ten thousand square meters. The underground city has been renovated, and there are plans to turn it into a modern tourist destination with restaurants, residential areas, shopping, and leisure centers throughout.[44]

The London Lowdown

London seeks to transform its abandoned tube stations into a powerhouse of urban development and address the problem of expansion in one of the most expensive cities in the world. The Old London Underground Company has plans to develop twenty-six sites with an estimated value of £3.6 billion. It hopes to convert them into retail parks, entertainment centers, offices, and cultural venues.[45]

Unfortunately, it seems that the project has been gripped and stalled by legal rows that may take years to adjudicate.

Bakersfield Below

In the early 1900s a Sicilian immigrant, Baldassare Forestiere, began turning what was useless farmland into a vast network of rooms, tunnels, and courtyards as a subterranean escape from the sweltering California Central Valley summer heat. Using only shovels, picks, and other hand tools, he was inspired to excavate for forty years, going as deep as twenty-five feet underground and spanning over ten acres, now known as Forestiere Gardens. He grew fruit trees and grapevines underground with natural light from skylights. Now guests from around the world tour through his grottoes and passageways.[46]

An Underground Stroll in Seoul

In 2017 Dominique Perrault's Architecture was selected as the winner of an international competition to design a multimodal hub and shopping center, the Gangnam International Transit Center, to be located in the heart of Seoul, South Korea. Included will be a multimodal hub, train station, urban park, and commercial complex, requiring large-scale underground construction in the city. In the design, there is a green network that weaves all streets and plots together and a continuous tree canopy in a new major landmark park offered to all habitants.[47] The ($1.15 billion) project to build a mammoth underground public transit terminal in Gangnam is planned to open by 2023.[48]

☼ ☼ ☼

Living beneath the surface: whether on land, in the sea, or on other planets, it is a slam dunk. It's an alternative that is happening and will continue to happen in the future. So keep looking down.

4

GOING BACK TO THE SEA

Back to Our Beginnings

There's nothing wrong with enjoying looking at the surface of the ocean itself. . . . Staying on the surface all the time is like going to the circus and staring at the outside of the tent.

—Dave Barry

YOUR CUP OF SEA?

We know more about space and setting up shop there than we know about setting up shop under the ocean. We know more about the topography of our moon and Mars and have better maps of those two celestial bodies than we do about the floors of our own bodies of water.

More humans have spent more continuous time in space than underwater. Up to now only a handful of humans have lived "down and under" for a short while. In fact, the record for living underwater in a structure (not a submarine) is held by two biologists who spent nearly seventy days in an undersea environment. Astronaut Peggy Whitson set the NASA record for most logged time in space—more than 650 days. Her main problem was lack of gravity, an aquanaut's would be lack of sunlight. [1]

THE THREE ZONES

The ocean is divided into three zones based on depth and light level. The upper 656 feet of the ocean is called the euphotic or "sunlight" zone. This zone contains the vast majority of commercial fisheries and is home to many protected marine mammals and sea turtles. Humans can't see much below 100 feet, but underwater lighting enhances their view while also attracting marine life.

The next zone is the dysphotic or "twilight zone." Some sunlight reaches this zone but not enough for photosynthesis to occur. This zone goes down to about 3,300 feet. The last zone is the aphotic or "midnight zone." No sunlight penetrates this zone, and it can reach depths of close to 20,000 feet. Sometimes people divide the midnight zone into two zones: the aphotic zone and the spooky sounding "abyss." At depths of 3,000–6,000 meters (9,800–19,700 ft), this zone remains in perpetual darkness. It alone makes up over 83 percent of the ocean.

Geographers divide the ocean into five major basins: Pacific, Atlantic, Indian, Arctic, and Southern. Smaller ocean regions such as the Mediterranean Sea, Gulf of Mexico, and Bay of Bengal are called seas, gulfs, and bays. Inland bodies of saltwater such as the Caspian Sea and the Great Salt Lake are distinct from the world's oceans. [2]

SCUBA MAN

Jacques-Yves Cousteau, coinventor of the self-contained underwater breathing apparatus (SCUBA), captured the imagination of the world by sharing his deep dives, exploration of shipwrecks, and discovery of previously unknown marine flora and fauna, especially through *The Undersea World of Jacques Cousteau*, his TV program from 1966 to 1976. At the time, Cousteau created a tidal wave of interest in undersea exploration. However, since then, interest in sending humans to live underwater for extended periods of time has ebbed. Once there were as many as sixty underwater habitats around the globe. [3] But excitement for such research habitats dwindled and the money dried up. The US Navy tried to staff SEALAB, an undersea experiment for itself and NASA in 1964. Unfortunately, SEALAB was discontinued after the death of an aquanaut who was trying to repair leaks in the structure. [4]

As of 2012 just three underwater laboratories remained, [5] as opposed to underwater hotels. At one of these, the Jules' Undersea Lodge, guests must actually scuba dive twenty-one feet beneath the surface of the sea. The lodge offers everything from education and training facilities to undersea weddings and luxury romantic getaways.[6] Jules' is in the Florida Keys, as are the Marine Lab, which is used as a research and training base by the US Navy, among others, and NASA's Aquarius, which NOAA plans to kill even though its budget is called "decimal dust," a little over $1 million.[7]

But the tide may be turning as private investors are starting to view the oceans as alternative places to build cities. Proponents and private investors maintain it could help alleviate overpopulation problems or offer a way to preserve our species—a kind of reverse Noah's Ark—in the event of a catastrophic cataclysm.[8]

LIVING DOWN BELOW THE WAVES

Making habitats with multiple modules made of steel, glass, and special cement for underwater use would be simpler than trying to create one giant bubble. Smaller structures could be added or taken away to create living space for as many people as desired. Most likely, we wouldn't want to build any deeper than one thousand feet, because the pressures at such depths would require very thick walls on-site and excessive periods of decompression for those returning to the surface.[9]

Some scientists believe living underwater is a logical solution to the population bomb or environmental collapse since it would be cheaper and easier to pull off than founding space colonies. And oceanographers believe that the only way to really understand what's happening in the ocean is to go down there for extended periods of hands-on time instead of just for an hour or so with SCUBA gear.[10]

The air needed to sustain aquanauts depends upon the depth of the habitat. Air could be either pumped from the surface or refiltered from down below via a chemical product called Sodasorb added to react with, and remove, carbon dioxide. Future aquanauts could also use artificially created life-forms or artificial "gills" to filter and/or harvest oxygen.[11] Currently, artificial gills are still science fiction, but companies are working on making them a reality and are in the very early stages of research and development.[12]

HOMESTEADING THE SEAS

Creating permanent dwellings at sea is called seasteading and has long been the stuff of science fiction, like the postapocalyptic movie *Waterworld*.[13] Academics and architects from the Seasteading Institute, cofounded by PayPal founder Peter Thiel and political economic theorist Patri Friedman, have created designs for permanent, innovative communities floating at sea and are working on prototypes for construction in the next decade. Also a think tank, the Seasteading Institute has been sponsoring studies on additional possibilities for ocean-based structures.[14]

Seasteaders are a diverse global team of marine biologists, nautical engineers, aquaculture farmers, medical researchers, investors, environmentalists, and artists. They plan to build floating islands, or seasteads, to host aquaculture farms, floating health-care facilities, medical research islands, and sustainable energy powerhouses.

Freedom from taxes, as fantasized about, would be a tricky question. The United States already demands that its citizens pay income tax even when they are living abroad—and seasteads would be considered "abroad." Until seasteaders are able to bank their money with independent, oceangoing financial institutions, they may not be able to escape the tax collector's clutches.

Some seasteaders think the way forward is to build less ambitious offshore communities to demonstrate the potential of the idea. By basing themselves just outside countries' territorial waters to avoid some of their laws, floating habitats could show land-based governments how such things as low taxes, light business regulation, and free access for foreign workers can produce wealth without ill effects.[15] But such exocommunities would have a hard time protecting their space—the US armed services might not come to their aid against modern-day pirates or internal disorder. And if giant corporations develop seasteads, they might take on the attitude and priorities of oligarchs with a "company-run town" attitude.

SEASTEADERS

Seasteaders who are thinking small for potential offshore communities are leaning toward prefab structures towed in segments—with floatation chambers so they won't sink—and then joined together at selected locations.

A myriad of new and old, reliable materials would be used to construct seastead structures: corrosion-resistant steel, lightweight, low-permeability foam cement, recycled plastic, and carbon fiber will be used to make modules. These structures could be added or deleted to create living spaces for as many people as desired while, again, going no deeper than one thousand feet.

Barge-like floating pontoon structures and platforms like offshore oil rigs built on floating columns are the most rugged possibilities for foundations but are also very expensive. Various shipbuilders have proposed an assortment of designs for floating cities based on massive "megafloat" pontoons, with skyscrapers towering above the waterline. Unfortunately, these would only work in calm, shallow waters—and would tend to be within land-based governments' territorial limits.[16]

Basic pontoon barge-type structures are the cheapest of options, but they are even more vulnerable than ships in choppy seas. In waters less than eight thousand feet deep, an option would be to moor a platform to the seabed, for example, to a number of barely submerged islands off the coast of California.[17] That said, in volcano and earthquake country, such islands might be available for only a short and scary stay.

British designer Phil Pauley has developed a concept for a sea habitat comprising interconnected spherical modules that could submerge to hide during storms and rest at the surface in good weather. To reach much deeper waters, communities could float freely or at anchor.[18]

Others are investigating this technique on a smaller scale. Do-it-yourself "ball stead" homes, for example, achieve buoyancy and stable surface using a heavy weight anchored well below the surface that keeps the ball-shaped structure from moving amid the waves.[19]

Simple cement structures reinforced with steel can displace massive amounts of water and last for decades or even centuries. For example, even after two thousand years of harsh beatings by the sea, a Roman harbor built with a mixture of standard concrete and volcanic ash is still intact in Pozzuoli Bay near Naples, Italy.[20]

The first floating city is expected to take to the water around 2020. Future cities built from scratch will be more dynamic, energy-efficient, and flexible. There is already research being done on ways to farm and harvest food and energy in deeper, more remote parts of the ocean. These cities of the sea could use algae for the production and storage of energy from wind and solar sources.[21]

THE LONG ARM OF THE LAW

The technical challenges are daunting enough, but the legal questions that seasteads would face, touched upon earlier, are no less tricky and call into question whether it would really be possible to create any kind of self-governing "mini-states" or "exo-states."

Many seasteaders who want to cut ties with the mainland will have to build their settlements anywhere from twelve up to twenty-four nautical miles offshore and regulate some economic activities in a two-hun-dred-mile "exclusive economic zone."[22] And a seastead tethered to a seabed (continental "land") would not qualify. So those gambling on "gaming," marijuana farms, prostitution, money laundering, and other dubious dealings prohibited on the mainland better give a second thought to disreputable and other piratical dealings at sea.

The United States, among other countries, asserts the right to ex-tend its jurisdiction in matters affecting its citizens across the entire planet. And like any other seagoing structure, a seastead would be obliged to register with a "flag state" to whose maritime laws it would be subject.[23] If the United States disapproved of any hanky-panky aboard a seastead, it could lean on the community by getting physical but prob-ably by first cutting off their banking, commerce, and supplies.

POLYNESIA BLUE FRONTIERS

The Seasteading Institute signed a Memorandum of Understanding with French Polynesia, creating a pathway for the first pilot project of floating islands governed under a "special governing framework."[24] The vision of the Blue Frontiers is to facilitate the development of conscien-tious and balanced settlements at sea where humans can peacefully coexist with the environment and with each other. The deal with French Polynesia specifies two points the project must prove before it gets the green light from the government: whether it will benefit the local economy and whether it will be environmentally sound. The pro-ject plan is to use sheltered waters not out in the open ocean, because while it's technologically possible, it is economically outrageous to go out to sea.[25]

AN ELECTRIFIED IDEA

If you need to build a structure—a hull, a breakwater, or even an entire island of your own—and you happen to live near salt water, you may soon be able to grow your own. The "homegrown" structure will be strong, durable, and should it ever fracture, it will be able to heal itself. The concept behind this is innovative and simple, and its plans are free.

Wolf Hilbertz, the originator of the concept, didn't want it to be commercially exploited, so he didn't file a patent. His tools were a pair of commercial garage battery chargers, some cable, and enough wire mesh fencing to make a pattern. He attached the cathodes (the charger clamps) to the wire mesh and found that minerals from the seawater (mainly calcium carbonate compounds—the same stuff as coral) were attracted onto the metal, making it look as if it had been dipped in molten glass and then, as it thickened, more like sprayed-on concrete.

Samples of the substance revealed that the material was able to withstand pressures of more than four thousand pounds per square inch, making it stronger than concrete but far lighter. Fish and other sea creatures, far from being repelled, were actually attracted to the mild electrical field too.

Hilbertz found that both the growth rate and the strength of the material could be regulated by adjusting the spacing of the anodes and cathodes and varying the density of the current. For power he used small, inexpensive wind plants for recharging the current.

This process could be used to construct buildings of any size, and the undersea supply of raw materials is estimated at about sixty quadrillion (that's fifteen zeros) tons of minerals. A by-product is pure hydrogen gas that bubbles up from the project. Experiments suggest that it might be possible to collect the gas so that the hydrogen could be used in marine cities or piped ashore for the generation of energy.[26]

DEEP SIGHTSEEING

A Polish company, Deep Ocean Technology, thinks tourism is the way to make seasteading economical. It has signed deals and is planning for multiple underwater hotels. It is also designing and constructing the world's largest underwater hotel planned for Dubai, with rooms both above and below the seabed.[27]

The Japanese are convinced that underwater cities could become a reality by 2030. A Japanese company's Ocean Spiral is a design consisting of a two-and-a-half-mile-long spherical city that is five hundred meters in diameter, within which a tower accommodates homes and workspaces complete with all needed resources and transportation that would accommodate five thousand inhabitants.[28]

US Submarine Structures, a company that specializes in creating underwater facilities and buildings, is now offering a very particular service that will allow anyone with the money to live in a pressurized house that sits on the seabed at depths of around sixty feet. The houses contain two floors, more than three hundred square feet of space, two bedrooms, a lounge, dining room, and even special feeding mechanisms to attract wildlife, providing a beautiful marine view and fresh fish. The hitch is that this spectacular home comes at an equally spectacular price of around $10 million. [29]

Another example of an on-the-water habitat is EcoFloLife's floating "WaterNest," an ecofriendly oval house. The approximately 330-square-foot unit is made of 98 percent recycled materials. Skylights, balconies, and large windows encircle the dwelling, allowing for efficient lighting and beautiful waterfront views. Almost 200 square feet of photovoltaic panels are on the rooftop, providing electricity. A sophisticated system of natural microventilation and air conditioning classifies it as a low-energy-consumption residential habitat. The units can be positioned on rivers, lakes, bays, atolls, and calm sea areas. The interiors are pleasant and include a living room, dining room, bedroom, kitchen, and bathroom. The layout can be configured as a house (for one to four people) or as an office, lounge bar, restaurant, shop, or exhibition space. [30]

Architect Vincent Callebaut and colleagues have proposed 3D-printed underwater "oceanscrapers" made from rubbish removed from the ocean. The concept is intended to highlight the dwindling natural resources on land and the need to clean up the "disgusting soup of petroleum-based waste" created by dumping plastic in the ocean.

Callebaut's domed marine structures measuring 1,640 inches in diameter would spiral down 3,280 feet from the surface and be designed to accommodate twenty thousand aquanauts. Food would come in the form of marine animals, seaweed, and farmed algae, while orchards and vegetable gardens would be grown on top of the domes.

The structures would be towed in prefabricated segments and then joined together and anchored at selected locations. Their internal construction would include floatation chambers that would render them practically unsinkable. They would be self-maintained and fully automated. [31]

DeltaSync has proposed a modular building strategy that would have movable parts. Its sea habitat would comprise interconnected spherical modules that could submerge during storms and float on the surface in good weather.

The habitat would be composed of modular platforms, either fifty-by-fifty-meter squares or pentagons with fifty-meter sides that could be arranged in branch-like structures. Concrete modules would support three-story buildings with terraces that could be used as residences, office and retail space, or hotels. [32] Sustainable features like hydroponic growing systems, biofuel production through floating algae, and protected fish and seafood habitats will result in "cyclical metabolism," thus making the floating ecosystem altogether more logical. [33]

UNDER SEA PREPPERS

Future Living Report, a website showing what housing will be like in one hundred years, suggests that we'll be living underwater because land space in urban areas will be in short supply or, worse, uninhabitable. There are those who see underwater living as a way of preserving our species in the event of an apocalyptic catastrophe, a kind of Noah's Ark.[34]

At the turn of the millennium, many parts of the world could look dramatically different—perhaps even underwater—as ocean levels rise. Rather than fight a losing battle against the tide, some predict that we will give in and live a life on or under the water.

That's the approach behind the Water-Scraper, the brainchild of architect and futurist Sarly Adre Sarkum. It is a futuristic, self-sufficient, floating city that looks a little like a giant jellyfish, a full-fledged underwater high-rise that harvests renewable energy, grows its own food, and even cultivates small forests on "green roofs." The buildings are kept upright using a system of ballasts aided by a set of squid-like tentacles that grip the bottom and also generate kinetic energy.[35]

CITIES OF THE SEA

A proposed manta ray–shaped vessel, *City of Meriens* (in English, "oceanite"), the creation of renowned French architect Jacques Rougerie, is a colossal floating university. It is capable of carrying up to seven thousand students, researchers, and professors on a four-hundred-foot vessel that runs completely on renewable energy, producing no waste. The ship also sports aquaculture facilities and a hydroponic greenhouse to support food production.[36]

Another idea is Blueseed, a Silicon Valley–based start-up company and seasteading venture on a converted cruise liner. The project is stationed far enough outside of territorial waters to hire engineers and scientists without the government red tape of getting work visas for foreign scientists.[37] The project has been floating since February 2, 2019, and is carrying out plans for a world of more competitive governance and greater ocean environmental health.[38]

Ocean Spiral

Shimizu, a Japanese corporation, and the University of Tokyo think that their technology for a submerged city will be ready within the next fifteen years, and the whole underwater world would take just five years to construct.

The "Ocean Spiral" could help coastal cities in the face of climate change and be completely sustainable with fish farms for food, a desalination plant for drinking water, and alternative forms of energy.

This structure, 1,500 feet in diameter, will be connected to a nine-mile-long path that winds through the sea toward a building located on the ocean floor, some 3,000–5,000 meters below the surface, containing houses, hotels, and business and commercial centers. It would also serve as a hub for scientists to mine natural resources and to examine possible ways to extract energy resources from the seabed.[39]

Green Float

The Shimizu Corporation also claims we have lost touch with what's really important and what truly makes us happy: healthy living, cultural pursuits, and contact with nature. It wants to reshape the cities of the future to help us reconnect with a healthy, happy lifestyle in an organic way. Its environmental city concept will have a waterbound base with a top that extends into the sky—overall resembling a natural plant. Residential space both at the waterfront and at the top of the tower will house forty thousand people per "island," while the tower will provide enough commercial space for ten thousand people to work. Island communities will be joined together in modules, making it possible for entire self-sustaining, carbon-negative cities to be built from groups of the floating platforms.[40]

Venus on a Half Shell

According to the Venus project's creators, it's an organization that proposes a feasible plan of action for social change, one that works toward a peaceful and sustainable global civilization. It outlines an alternative to strive toward, where human rights are no longer paper proclamations but a way of life.

A civilization comprising a global network of cities in the sea could easily accommodate many millions of people and relieve land-based population pressures. Some cities could also be used as a new resource for mining the relatively untapped resources of the oceans without disturbing their ecology. Still others might monitor and maintain environmental equilibrium and reclaim dangerous radioactive and other pollutant materials that have been dumped into the sea. Such systems could be used to cultivate and raise fish and other forms of marine life to help meet the nutritional needs of their communities' people.

As in other innovative cities, some structures could be towed in prefabricated segments, joined together at selected locations, and anchored to the ocean floor.

A central dome or theme center would contain educational facilities, computerized communications, networking systems, and health and child care facilities. Three rings of buildings adjacent to the center would house the research facilities. There are also community centers for cultural activities, dining, and other amenities.

Eight residential districts would have a variety of unique, free-form architecture. Each home would contain gardens and would be isolated from other homes via landscaping. Indoor agricultural hydroponic, aeroponic, and aquaponic facilities and outdoor agricultural belts would be used to grow a wide variety of organic plants without the use of toxic chemicals. Areas would also be set aside for clean, renewable sources of energy such as wind generators, solar concentrators, and geothermal, photovoltaic, and other such systems.

Such prefabricated, modular homes are built of a new type of pre-stressed, reinforced concrete with a flexible ceramic external coating that would be relatively maintenance-free, fireproof, and impervious to the weather. Their thin-shell construction can be mass-produced in a matter of hours.[41]

Aequorea

Another of Vincent Callebaut's concepts is to build a futuristic skyscraper city nearly a mile below the water's surface. It calls for a series of domed buildings with 1,000 towers and 250 floors—all made from 3D-printed plastic from ocean waste—that would house 20,000 people. Its jellyfish-like geometry would allow it to remain unaffected by high currents, storms, earthquakes, and other natural water movements.

Bioluminescence would light the buildings. Underwater people would use gill masks (a device that extracts oxygen from water and dissipates carbon dioxide). It includes a transport system powered by seaweed.[42]

Bucky's Triton

In the 1960s, futurist Buckminster Fuller and staff designed a floating city, Triton, a concept for an anchored floating city for one hundred thousand people that would be located just offshore and connected with bridges to the mainland. Its plans even included a geodesic ring of spherical modules. Fuller's SubBiosphere2 would float in fair weather and then submerge whenever the seas became rough. The design was prepared for the US Department of Housing and Urban Development.[43]

Blue Revolution Hawaii

Blue Revolution Hawaii, a combined US and Japanese project, is another organization formed to educate the public about planning for a future with thousands of floating cities built with the resources from the ocean. According to its creators, the oceans offer the most opportunity as our next frontier. This program can be accomplished with one-tenth of the cost of the $150 billion International Space Station while developing sustainable resources and enhancing the environment.[44]

Syph

The proposed undersea city of Syph, in Australia, would evolve into a collection of living spaces with specialized functions like energy generation and sustainable food production working together.[45] Far from being impractical utopias, Syph as well as other undersea or floating cities could be every bit as integrated into global society as the ones we already have on land.

FLOATING SOME H₂OMES

Apparently there are plenty of big, secondhand liners selling cheaply. Ship-shaped structures can pack in more apartments and office space for a given cost than other types of floating design, but they have a big drawback because of their tendency to roll in choppy seas. Cruise ships can sail around storms, but static seasteads need to be able to ride them out. And the stabilizers on big cruisers only work in moderate seas and when the ship is moving.

A full-service floating city already exists for residents of *The World*, a 644-foot yacht that continuously circles the planet. Launched in 2002, the ship contains 165 condominium spaces that sell for millions.

Nicknamed the *Freedom Ship*, the largest private residential ship on the planet is essentially a mile-long flat-bottomed barge with a high-rise building on top. Weighing three million tons and with a top speed of ten knots, the floating city would circle the globe every three years, stopping twelve miles offshore at each port for a week at a time. High-speed ferries would connect the forty thousand residents and twenty thousand crew members to the mainland and bring back visitors. "We won't just be visiting those countries," says *Freedom Ship* director and executive vice president Roger Gooch. "We anticipate those countries visiting us."

The *Freedom Ship*'s size—and its $11 billion price tag and three-year construction process—stretches credulity. Too big for any existing shipyard to build, the ship must be constructed in pieces and—a familiar idea by now—towed out to be assembled at sea.[46]

A floating village at London's Royal Docks has the official nod, and Rotterdam has a Rijnhaven waterfront development experiment well underway. Eventually, whole neighborhoods of water-threatened land could be given over to the seas. After decades of speculation and small-scale applications, the floating solution is finally enjoying political momentum—and serious investment.

Slums to Minicities

Architect Kunlé Adeyemi proposes a series of A-frame floating houses to replace existing slums on waterfronts. As proof of concept, his team constructed a floating school for the community. Still, many buildings do not a city make: infrastructure remains a problem. One solution would be to use docking stations with centralized services, rather like hooking up a caravan to power, water, and drainage lines at a campground.[47]

Floating communities would allow people to continuously move around to take advantage of the best climates or to avoid serious storms.

STRENGTH OF SEAGOING POWER

Addressing energy is still in the planning stages for large structures. Sustainable future options include harnessing wave, tidal, and current action, or placing solar panels on surface barges.[48]

The heat locked in Earth's vast oceans could generate useful energy—the basic thinking behind OTEC (ocean thermal energy conversion). According to some estimates, there's enough heat in the upper layers of the oceans to meet humankind's energy needs hundreds of times over.[49]

The abundant wind and sun available at sea could power turbines, OTEC could harness the temperature difference between the surface and the depths—a process that also provides freshwater as a by-product—and CO_2 could be captured from the surface and then converted into methane with the help of microbes called methanogens that could provide vast amounts of energy.[50] This power will be used to monitor life support systems with coordination from smart surface automation.

LOTS OF OMEGA-3 AND WATER, WATER EVERYWHERE

Fresh seafood is generally easy to come by in the ocean. Aquanauts could cultivate and raise fish and other forms of marine life and, in the future, help meet the nutritional needs of the world above. Supplying water would be no problem via desalination processes, including reverse osmosis.[51]

The drawbacks of current fish farming have created opportunities for technology. For example, floating "drifter pens" can replace stagnant ponds with GPS-tracked cages stitched out of copper wire to enable a constant inflow of fresh ocean water without losing the precious fish. Such geodesic aquariums will be let loose in swirling ocean gyres, where they will maintain a position so vessels can meet them and pick up their "harvested" goods.[52]

MINES AND MINERALS

There is a lot of interest in deep-sea mining and exploration. The Chinese in particular have been investing in deep-sea expeditions to investigate the possibility of mining manganese nodules, rocks that contain nickel, copper, cobalt, manganese, gold, and other valuable, rare earth minerals. And many marine biologists think that the best way to explore the ocean is by living there rather than visiting.[53]

FIRST FLOATING NATION

The world's first floating nation is planned for the Pacific Ocean by 2020, to be built by the nonprofit Seasteading Institute. The radical, futuristic plan, bankrolled by PayPal founder Peter Thiel, calls for an eventual floating nation with networks of hotels, homes, offices, restaurants, and more. The floating islands of the nation will feature aquaculture farms, health care, medical research facilities, and sustainable energy powerhouses. Cities will be built on a network of eleven rectangular and five-sided platforms so they can be rearranged according to inhabitants' needs. The square and pentagonal platforms will measure 164 feet in length and they will have 164-foot-tall sides to protect buildings and residents.

The platforms, accommodating between 250 and 300 people, will be made from reinforced concrete and will support three-story buildings where apartments, terraces, offices, and hotels will be located. Hotels will have "green roofs" covered with vegetation. Construction will use local bamboo, coconut fiber, wood, and recycled metal and plastic and will last for up to one hundred years, according to the plans.

The Seasteading Institute hopes to "liberate humanity from politicians." The radical plans call for networks of communities floating in international waters and operating within their own laws. The president of the Seasteading Institute has said he wants to see "thousands" of rogue floating cities by 2050, each of them "offering different ways of governance." The institute hopes to raise around $60 million by 2020 to build a dozen buildings; ultimately a total of $167 million will be needed. [54]

Many of these structures and cities may serve as oceanographic universities whose efforts include maintaining the ecological balance of marine systems, maintaining sea farms that raise fish and other forms of marine life to help meet the nutritional needs of the world's people, and cultivating marine life in general. Other anticipated prospects include mining the oceans without disturbing their ecosystems, monitoring and maintaining environmental equilibrium in general, and reclaiming dangerous radioactive and other materials that have been dumped into the sea.

The oceans are our birthplace, although we haven't treated them as such. We should give a lot of thought to the idea that they may be one of our last refuges for survival on an increasingly hostile terra firma.

5

SETTLEMENTS ABOVE THE SKY

Cities in Space

The solar system is to us what the American West was to Lewis and Clark, and, showing the same vision and courage they once did, we should go forth boldly into the new frontier.
— Jeffrey E. Brooks, "Jefferson's Dream"

A TREATY FOR SPACE

First, a cautionary word before we take another "giant leap for [hu]mankind": the Outer Space Treaty was formed by the United Nations and entered into by the United States, the United Kingdom, and the Soviet Union on October 10, 1967.[1] As of July 2017, 107 countries were parties to the treaty, while another 23 have signed the treaty but have not completed ratification.

According to the treaty, all parties are prohibited from placing nuclear arms or other weapons of mass destruction in orbit, on the moon, or on other bodies in space. Nations cannot claim sovereignty over the moon or other celestial bodies.[2]

Despite this, US president Donald Trump signed an executive order in December 2018 for the creation of a "US Space Command."[3] The United States Space Force (USSF) is the proposed space warfare service branch of the US armed forces and is intended to have control over military space operations. It would be the sixth branch of the US armed forces and the eighth American uniformed service.[4]

Although it is supposedly used to "consolidate space operations under a single authority," according to CNN it could also be a response to what news outlets have called a "space arms race," as powers like China and Russia are developing possible antisatellite technology.[5]

The new space command won't be the first one founded by the United States. One was established in 1985 but disbanded in 2002 in the aftermath of the 9/11 terrorist attacks, when attention was refocused on homeland security.[6]

Judging from Trump's administration, the president's flight of fancy would seem to be almost certainly seeking to establish military superiority in the heavens. Per the Outer Space International Space Treaty, Trump won't be allowed to weaponize space—although nothing has stopped him in the past from fracturing treaties.

Hopefully Trump's aggressive pitch is just a cartoonish form of jingoism. But his hunting dog, Mike Pence, rattled his saber, too, stating that plans are in place for forming an "elite group of war fighters specializing in the domain of space."[7]

Next, a word about the exploration and the commercial exploitation of space and other celestial bodies: The examination and use of outer space shall be carried out for the benefit and in the interests of all countries and shall be the province of all humankind, according to the outer space treaty, specifically:

- Outer space shall be free for exploration and use by all states.
- Outer space is not subject to national appropriation by claim of sovereignty, by means of use or occupation, or by any other means.
- States shall not place nuclear weapons or other weapons of mass destruction in orbit or on celestial bodies or station them in outer space in any other manner.
- The moon and other celestial bodies shall be used exclusively for peaceful purposes.
- Astronauts shall be regarded as the envoys of humankind.
- States shall be responsible for national space activities, whether carried out by governmental or by nongovernmental entities.
- States shall be liable for damage caused by their space objects.
- States shall avoid harmful contamination of space and celestial bodies.[8]

Here's a good question: Will the "development" of the planets become a billionaire boys' club, or will it be sucked up and/or manipulated by corporate conglomerates—all seeking planets for profit? Despite progress in technology and the allure of valuable resources, space settlement has been hampered by the lack of clearly defined legal rules for recognizing property rights in space under current United States and international law.[9]

There is internationally recognized legal precedent for retaining ownership of resources mined in space, as lunar samples have been returned to Earth on both the United States' and Russia's behalf. But actually owning the portion of the celestial body from which the resources are harvested—as in a traditional mining claim—is more problematic. Without legally recognized rights to buy, own, and sell titled property, it is difficult if not impossible to raise capital to develop land or extract the resources it holds.[10]

But someone or some entity will find a way. A wily entrepreneur has already been selling extraterrestrial real estate. In 1980 the (US) Lunar Embassy Commission started selling real estate plots on the moon. As of 2009 it claimed to have sold 2.5 million one-acre plots for around twenty dollars per acre. The legal loophole the company crawled through was the contention that the law prohibits nations, not individuals, from selling chunks of green cheese.[11]

However, without exploration and exploitation of those resources, there may not be any civilization in space, as large-scale colonization of space is going to take very big bucks. The monetization of the moon, Mars, and space can't be claim-staked only for plans of profit but has to be executed with careful preparation, protections, and protocols. Unfortunately, in the past we have seen profits supercede the rights and the property of others.

But as the beginning of the race to space spurts, profits will push the pursuit for the treasures of outer space. Even in *Mars*, the fictitious TV series, after a short time on the planet corporate visitations almost immediately cause concerns and arguments about rights to precious resources, like water.[12] Will the brave new world of outer space continue to be business as usual?

THE FINAL FRONTIER

It took humankind about two hundred thousand years to colonize almost every corner of Earth[13]—while always looking up at the stars—and less than a century from propeller to propulsion to leave our planet as we begin to inhabit space.

There are many reasons for venturing out into the ink and vacuum of space, from insurance against extinction to scientific inquiry, enjoyment of travel, and—perhaps the most compelling—the human compulsion to see what's beyond the next horizon. Gene Roddenberry wrote of striving "to boldly go where no man has gone before."[14] Stephen Hawking said, "It is time to explore other solar systems. Spreading out may be the only thing that saves us from ourselves."[15]

It should be pointed out that not everyone is in agreement. "This whole idea of terraforming Mars—are you guys high?" Bill Nye the Science Guy sputtered in an interview with *USA Today*. "We can't even take care of this planet, let alone another planet," added his pal, respected astrophysicist Neil deGrasse Tyson.[16] Tyson believes any plans to make Mars livable for humans are absurd, but he does believe we should send astronauts to explore the Red Rock.[17]

SETTLEMENTS IN SPACE

A space settlement is a structure like a small city in size and built either within Earth's orbit or on another planet. At present the International Space Station (ISS) currently houses only six astronauts at a time, but a space settlement would have hundreds or thousands, or maybe millions, on board.[18]

The four most common arguments in favor of colonization are survival of our civilization in the event of a planetary scale disaster—a kind of "get out of jail free" card—the availability of additional resources in space that could finance the expansion of humankind, simple conquest, and pure scientific exploration. The most common objections to colonization include the fact that the value of raw materials in the cosmos may likely be irresistible to major economic and military institutions[19] and provide a reason for conflict.

Needless to say, the building of a space colony will present a set of mammoth technological and economic challenges. Living in space and on other planets is going to be a tough techno-nut to crack. Besides the volatile and careening temperature changes, radiation, and lack of immediate resources, there're the horrendous vacuum of space, dangerous moon and Martian razor-shaped dust everywhere that contains carcinogenic chemicals,[20] and countless other obstacles to overcome on other planets. Unless we can crack open asteroids and the poles on Mars for water to ensure enough oxygen, people won't last months, let alone years.

However, high-tech ionizing and filtration systems could be used on the moon and Mars to keep habitats dust-free, and the water they bring, which will be constantly cleaned and reused, will be enhanced by Martian ice at the poles. Then there is the predicament of food and habitation. The use of human waste for fertilizer mixed with the questionable soil of Mars would help,[21] but so far we don't know what the food value and yield would be, although it looks promising.[22] And hydroponics, which is the best bet for growing produce, is still problematic due to water shortages. If burrowing underground or in caverns is not possible, robotic 3D printers might be able to make thick enough walls to shelter against radiation.[23]

Space settlements would have to provide for just about all their material needs in an environment that is hostile to humanity. They would require technologies for controlling life support systems that have yet to be developed. They would also have to deal with the physical and psychological issues of how to behave and thrive long-term in such places. Expenses would have to be reduced, as sending anything from the surface of the Earth into orbit costs around $2,500 per pound.[24] Of course, this would decrease with increases in technology—such as space elevators and platforms, space planes, and inexpensive, reusable launch systems—reducing the cost to about ten dollars per pound to Earth.[25]

That doesn't mean that there aren't plans in place by entrepreneurs, scientists, and several countries that are hot to trot for the first shot at an extraterrestrial colony. Just fifty years ago, after the moon landing, we were so infatuated with the idea of space travel that it seemed only decades away. The Space Shuttle was so named because it was intended to make fifty round trips per year, and there were active plans for expanding civilization into space, the moon, Mars, and beyond. But interest, passion, the Cold War, and economics sagged along with the deflated NASA budget that sank from 5 percent of the US federal budget to less than 0.5 percent.[26]

But now the children of the Apollo age have resurrected the hot-to-trots to take on some of the biggest difficulties of large-scale space settlements. On the celestial menu are basically a few choices: terraforming, using massive geoengineering projects to generate a new environment; creating biodomes to build an alien metropolis;[27] burrowing beneath the surface of Mars for protection and water;[28] helium-filled airships cruising through the Venusian atmosphere[29]—a colony in the clouds—and a low-Earth-orbit model expanding upon the current space station scheme of a "planar-cluster," housing billions of people across thousands of miles of space.[30]

Another possibility is an outer space model of floating cylinders with artificial gravity (think *2001: A Space Odyssey*); survival would be based on towing and digesting the natural resources of asteroids, which contain water and other raw materials. As the saying goes in space travel, "Once you're out of Earth's gravity you're halfway to anywhere."[31]

OUR SATELLITE STEPPING-STONE

There is no atmosphere and very little magnetic field to protect moon and Mars colonists from radiation, which is lethal and could render them cancer-ridden or sterile. However, with the proper protection both inside and outside, a transplanted "Martian" could live for sixty years before any problems would arise.[32] As far as the moon is concerned, at one time the pale rock was volcanic, and it has a number of ancient lava tubes left over that would be ideal for living spaces.[33]

A lunar city has the advantage of being practically next door to Earth and as such could participate in Earth's economy to some extent. Possible anchor industries could include space tourism and mining as well as industries that require microgravity.[34] The first rule for planning a base on the moon is that you want to use as much lunar material as possible, especially for 3D printing.

There have been rumors that the United States and Russia would be teaming up to build a lunar base. Russian Roscosmos and NASA have both released statements saying the two countries share a "common vision for human exploration" and will cooperate in building a "deep space getaway" starting in 2020.[35] This is a big deal for many reasons, as working together, despite earthbound political tensions, means that our countries might be less likely to start a shooting war in space.

THE RUN FOR THE RED PLANET

From books amd movies such as H. G. Wells's *War of the Worlds* and *The Martian* to the TV program *Mars*, the idea of extraterrestrial life or pioneering Mars has been a popular theme of sci-fi space culture. Possible methods to help make the planet habitable could be using bioengineering organisms to convert carbon dioxide in the atmosphere to oxygen or darkening the Martian polar caps to reduce the amount of sunlight they reflect to increase the surface temperature and provide water.[36]

Dr. James Green, director of NASA's Planetary Science Division, presented the extraordinary idea of putting a magnetic shield around Mars to restore its atmosphere. It would shield the planet and generate an artificial magnetic field, allowing Mars to slowly restore its atmosphere, deflect radiation, and perhaps deliver up part of its ancient oceans. The shield would also protect the planet from solar winds and the greenhouse effect.[37]

NASA has established three phases that must be completed before dropping off humans on Mars's doorstep. The first is one that NASA calls "Earth-reliant," in which scientists continue to test the feasibility of living in space and conduct more research aboard the ISS. The second, "proving ground," consists of operations around the moon to establish how humans will return to Earth safely. In the third, "Earth-independent," humans will establish a self-sufficient colony on Mars.[38]

When NASA gets to Mars, sometime in the 2030s, its rather optimistic goal will be a proper human settlement, first using a spacecraft capable of sustaining a crew in orbit around Mars, with accompanying landers.[39] SpaceX chief executive Elon Musk bets on landing an unstaffed spaceship by 2022 and having the first one hundred humans onboard for a settlement by 2024.[40]

Jeff Bezos, founder of both Amazon and the space company Blue Origins, hopes to set up a cargo delivery service to the moon in 2020. Bas Lansdorp of Mars One intends to send a group of humans to Mars—one way. Of 202,586 applicants, only 100 have been selected. Plans have been pushed back numerous times, and the most recent timeline indicates that Lansdorp hopes to launch a crew by 2031.[41]

The Sierra Nevada Corporation (SNC) has been given the go-ahead from NASA to begin full-scale production of its "Dream Chaser" commercial space cargo plane, scheduled to make its first mission in 2020. It's capable of carrying approximately 12,125 pounds of cargo and returning to Earth on a runway landing. Increased competition between corporations is helping to reduce the high cost of launches and is opening new doors for smaller companies.[42]

Another flag posed to be planted on Martian soil belongs to the United Arab Emirates. They have a tendency to think big and move slowly, and they are set to build the Mars Scientific City as part of the Mohammed Bin Rashid Space Centre's Emirates Mars Mission, which aims to establish a viable human colony on the planet within a hundred years.[43]

China is seeking to become the first country to conduct joint robotic orbital and surface exploration of Mars in a single mission by 2020. As for a potential partnership with China, since 2011 the US Congress has prohibited NASA from working with the country because of national security concerns.[44]

THE VENUS AND MARS VENUES

The idea of sending humans to Venus and Mars was first proposed after Neil Armstrong's historic moonwalk, when NASA was flush with funds and promise. Robotic explorations of Venus could potentially lead to the development of a human mission to explore the clouds of Venus by solar-powered aircraft that ultimately could even envision colonization in the Venusian atmosphere.[45]

If you like it hot, you'll love Venus—it's scorching enough to melt lead, and its acid rain will blister flesh off bones. Venus is the "cloud planet," nicknamed for its thick swaths of carbon dioxide (97 percent), sulfuric acid, nitrogen, and trace elements (3 percent). It's a gross example of a greenhouse-gas-affected planet.[46]

Venus is four to five months closer to Earth than Mars is. Opportunities of a trip to Venus come once every 1.6 years, compared to every two years for Mars. The gravity on Venus is about 90 percent that of Earth's, and Venus is 30 percent closer to the sun than we are. If Mars has almost no atmosphere, Venus has too much; it is ninety times thicker than Earth's. However, about thirty miles above the surface, the air pressure is close to Earth's and has enough atmosphere to shield radiation. The temperature at that height is almost the same as Earth's, or about 70°F.

Carl Sagan advocated bombing the upper atmosphere with genetically modified blue-green algae to reduce the carbon-dioxide-saturated atmosphere to a level conducive to supporting terrestrial life. However, three decades later Sagan himself declared the idea "fatally flawed."[47]

And since Venus's gravity is nearly as strong as Earth's, colonists living there wouldn't develop the brittle bones and weak muscles associated with low-gravity environments. Another happy coincidence is that CO_2 is heavier than air, so a balloon filled with oxygen is lighter than the Venusian air and will float in the sky with no problem, part of a plan called the High Altitude Venus Operational Concept (HAVOC).[48]

The scenario of a balloon deflating by ripping or popping open is not as probable as it simply slowly leaking, because the pressure inside the balloon would be the same as the pressure outside.

A dirigible one kilometer in diameter will lift seven hundred thousand tons; one with a two-kilometer diameter would lift six million tons and offer an environment as spacious as a typical Earth city. And dirigibles could be made out of Teflon, which is tough enough to protect against sulfuric acid clouds. Scientists are figuring that a Venusian settlement of one hundred people will require a slew of resources to keep its inhabitants alive. However, if the project works, there are plans for massive dirigibles for thousands of potential Venusians.[49]

The CO_2 atmosphere of Venus could be split into oxygen and carbon, and the sulfuric acid could be split into water, oxygen, and sulfur. And while the surface of Venus would remain inaccessible to humans, robots could explore and mine the rocky terrain, and there would be no need for complicated terraforming.[50]

Other bizarre, outlandish plans to pacify the atmosphere have included infusing the planet's atmosphere with forty quintillion kilograms of hydrogen obtained, somehow, from the gas giants Jupiter and Saturn, and removing the excess carbon dioxide in the atmosphere in a process called a Bosch reaction. This would produce enough water to cover 80 percent of the planet's surface.[51]

RIDING AN ASTEROID AND POSSIBLY OTHER PLANETS

A foothold in space might be had by snagging a trip on an asteroid. We could capture it, spin it to induce gravity on the interior, and freewheel a free ride in space, unbound by planetary gravity.[52]

The sun's energy will enable immense amounts of power for electricity, manufacturing, or mining. It's been estimated that a specific asteroid could contain more gold than has been mined on Earth.[53] Asteroids such as Ceres have enough surface area to provide uncrowded homes for more than a trillion people. Buildings could be so large that they would feel like the outdoors.[54]

Beyond the asteroid belt, another possible spot is Jupiter's moon Europa. It's completely covered in a thick layer of ice, protecting enormous oceans. So life might be thriving beneath this protective crust, and if so, it might be a welcome place for us.[55]

Still another idea is the Analemma Tower, a skyscraper that would be suspended from an asteroid placed into orbit right above earth, at the height of about thirty-one thousand miles. The advantage of making such a building is that since it's being built in the air, it can be constructed as large as needed and transported anywhere. It would be powered by solar panels constantly exposed to sunlight. Water would be recycled in a semiclosed loop.[56]

But before we blast off to extraterrestrial regions, we will need the help of city planners to execute where and how we are to survive and thrive as engineers and scientists develop life support systems; develop sources of food, water, and fuel; overcome the negative effects that living in space has on the body and mind; and find a faster way to travel.

As for gas giants like Jupiter, Saturn, Uranus, and Neptune, while there isn't a solid surface to build on, the planets might also sustain floating cities. But we would need to be heavily shielded from radiation, and technology for that is not available just yet.

These gas giants are also known as Jovian planets. It's unclear what the dividing line is between a rocky planet and a terrestrial planet and some may have a liquid surface. For example, in our solar system, gas giants are much bigger than terrestrial planets, and they have thick atmospheres full of hydrogen and helium. On Jupiter and Saturn, hydrogen and helium make up most of the planet, while on Uranus and Neptune, the elements make up just the outer envelope. These planets are also inhospitable to life as we know it, although this region of the solar system has icy moons that could have habitable oceans, but at present they are out of bounds for any type of human exploration or colonization.[57]

AIR, FOOD, AND WATER

When we choose to move into permanent space colonies, a new set of challenges will arise. A large one is sustenance. Our only colony outside of Earth is the ISS. It's only 254 miles from Earth and relies on a continuous resupply of resources for its crew of three to six people. NASA is working on developing techniques to regenerate oxygen from atmospheric by-products, such as the carbon dioxide we exhale.[58]

Mars can be fairly comfy (70 to 100°F) during the day in the warm season but very cold, especially at night (–90 to –200°F), so plants will have to be kept covered indoors. Its atmosphere of choice is primarily carbon dioxide, so we can grow plants there.[59] Agriculture will be essential to settling Mars, as every resupply mission will take up to ten months, and a suitable window for launching those missions only happens once every two years.[60] Scientists have raised crops in simulated Martian soil and found that it's surprisingly arable for carrots, green beans, tomatoes, and potatoes—except that they only grow to about the size of a salt shaker, and their taste is reportedly . . . blah.[61]

Bland foodstuffs made of algae called chlorella or spirulina might not look like four-star dishes, but they are loaded with powerful nutrients such as most of the necessary vitamins, minerals, iron, magnesium, beta-carotene, gamma-linolenic acid as well as all the essential eight amino acids. They are also rich in potassium, phosphorous, and calcium, and contains significantly large amounts of protein when compared to various other sources.[62] They contain sugars and proteins, but the balance between the two depends on the nutrients in the water, the light the algae receives, and how you can manipulate the taste—again . . .

blah. But with 3D printers we can produce everything from pizza to chocolate. (See chapter 10.)

An initial settlement would need to carry a certain amount of water and recycle all waste liquids. This is already done on the ISS, where no drop of liquid—washing, sweat, tears, and even urine—is wasted.[63] A colony would also likely try to extract water, possibly from underground supplies of liquid that may exist on Mars or from ice on the poles.[64]

On the ISS, oxygen is generated through a process known as electrolysis that separates it from the hydrogen in water, but the process only has 40 percent efficiency. We need to solve how to convert hydrogen or CO_2 directly into oxygen. Crops can also be used to convert carbon dioxide in the air back into breathable oxygen.[65]

ENERGY

Another requisite will be to develop ways to produce electricity. A substance called perchlorate, a human-made chemical used in the production of rocket fuel, missiles, fireworks, flares, and explosives exists in relatively high quantities on Mars, making up about 1 percent of all the dust on the red planet.[66]

The value and intensity of sunlight on the surface of Mars is lower than on the surface of Earth because it is farther from the sun. Mars's atmosphere is also subject to periodic sandstorms, which are notoriously problematic, limiting the amount of available light and increasing sand piles on solar panels.[67]

Fossil fuel stores will quickly become depleted unless we can construct supply stations to produce methane as a fuel. And it's obvious that future cities will need a cleaner, renewable energy source. An answer might be a "Luna Ring," a permanent array of solar collectors around a planet's equator. The majority of the solar cells would always face the sun and collect massive amounts of solar energy, which would then be beamed via microwave power transmission antennae.[68]

Another possibility being tested is the LightSail 2, which moves through space by harnessing the power of solar photons, no fuel tanks or thrusters required.[69]

HOMES FOR HUMANITY

Another requirement for habitats is an environment able to maintain air pressure, temperature, and protection from radiation. The first settlements would probably be made of domes, like Bucky Fuller's geodesic domes, because a sphere is the best way to hold in pressure and is structurally strong.

Structures will have to be built with what's available from boring machine by-products of rammed earth built by robots, or by second-generation, lightweight 3D printers using the soil of the planet as a building material. Missions to the moon and Mars have proved that there's silicon available for windows, aluminum, iron, and magnesium for parts of the main structure.[70]

The interiors would be basic, collapsible forms of 3D-printed materials to make the best use of available room. Think space-age Motel 6. It will be functional, not glamorous, which will come later when resorts and hotels are built for tourists.

Freewheeling space settlements must be airtight and should rotate to provide pseudo-gravity. People standing on the inside of the hull will feel gravity, but when on the outside, they will enjoy the fun and freedom of weightlessness. Settlements in low Earth orbit (LEO) would be 760 times closer than the moon and 100,000 times closer than Mars. The moon would be a few days away, but trips to Mars would still take many months.[71]

According to outer space entrepreneur Elon Musk, a city on Mars with a million inhabitants could be achievable within fifty years, complete with factories and ice cream and pizza parlors. He claims he will build a "Mars Colonial Fleet of more than one thousand cargo ships which could transport 200 passengers at a time, along with tons of supplies."[72]

Musk says, "Probes and robotic rovers have already been on or around Mars for 40 years. It would be quite fun to be on Mars because you would have gravity that is about 37 percent of that of Earth, so you would be able to lift heavy things and bound around." He adds that "he wants to be on the first flight to Mars, where he would like to die, but not on impact."[73]

NASA has also said it is planning to establish a Mars colony by the 2030s. But the agency has plans to establish a base on the moon first, to create "deep-space habitation facilities" that will act as stepping-stones to the red planet.[74]

TECH TOOLS AND MATERIALS

There will also be a weight limit for supplies, meaning colonists will have to watch their calories and rely largely on local resources for their building materials. Surprisingly, an effective way to get building materials is to pound the ground with a sledge hammer. Scientists found that this primitive method could be used for making Martian bricks—stronger than steel-reinforced concrete—by 3D printing Martian ice, turning it into vapor, then converting it into water, and using it to print solid structures. Construction would use digital manufacturing techniques and autonomous machines, releasing people for other tasks.[75]

THE ELEVATOR INTO SPACE

In the future people won't be ferried into space by huge booster rockets but via elevators made of grapheme carbon nanotubes. One hundred times stronger than steel and more flexible than muscle, these will be anchored to Earth by a tether and held under tension by centrifugal force via a counterweight thousands of miles high.

The concept of a space elevator is older than the first space film (c. 1902); the first space elevator was proposed in 1895. The present project calls for sending a robot two kilometers up via a cable and building a test platform of high-altitude balloons that are tethered to the ground. Then the robot launch could help with the lunar elevator, which in turn could help with the Earth elevator.

In the lunar version, a space capsule would be attached to a rocket and sent toward the moon. When it got close enough, a cable would eject from the capsule and attach to the surface of the moon, allowing for transport between the surface and the capsule hundreds of thousands of feet high.[76]

The biggest structure ever made might be built from one of the smallest materials ever made—nanotubes approximately fifty thousand times smaller than the width of a human hair. They are strings of extremely "sticky" carbon atoms that bind together, becoming stronger and stiffer than any other known material. The space elevator would be made by "cooking" carbon in a special way and aligning the nanotubes vertically—like trees in a carbon forest. So far, the longest nanotube made has been only a few millimeters long—a little short of the planned twenty-two miles. Scientists claim that it probably can be done—when the technology catches up with the concept.

Although the space-elevator concept was once thought to be the stuff of science fiction, some aerospace engineers believe the idea is essential to the future of space exploration as an alternative to building ever-larger rockets; in terms of payload, rocket power has more or less reached its limitations.[77]

"Cubesats," are miniature satellites that have been used exclusively in low Earth orbit for fifteen years and are now being used for interplanetary missions. In the beginning, they were commonly used in low Earth orbit for applications such as remote sensing or communications. In the future they would be released onto the elevator as it moves along a cable with its own electrical, solar-powered motor. Robotic climbers would patrol up and down the tether, reporting when to replace parts.[78]

SPACE MOTELS WITH TV

Today the biggest space operation isn't NASA, Musk, or even the US Defense Department, but DirecTV, which is very interested in low-Earth-orbit transmission, called New Space, a $50 billion business.[79]

Space capitalists include hotel tycoon Robert Bigelow, the owner of Budget Suites of America. He founded Bigelow Aerospace in 2015 and was supposed to build a "commercial space station" in 2015, but apparently it's been a little slow on the takeoff.[80]

Bigelow plans to use a SpaceX rocket to send one of his inflatable space habitat modules up for testing at the ISS. These blow-up houses are capable of operating independently as space stations, and the motel magnate wants to lease them as suites, laboratories, or maybe lunar motels.[81]

SIGHTSEEING IN SPACE

The Space Tourism Society (STS) was founded in 1996 by Jim Spencer as the world's first society focused on the establishment and expansion of a profitable space tourism industry. One of the Society's long-term research, design, and promotional programs is establishing an orbital super-yachting community complete with snuggle tunnels for zero-gravity sex.[82]

Space tourism is no longer a distant dream as multiple companies want to be the first to get tourists off the ground for a once-in-a-lifetime vacation. Well-heeled tourists will soon pay big time to go on "cruise ships" either in LEO or a nice, week-long trip around the moon. Orion Span joined the ranks of Virgin Galactic, Axiom Space, SpaceX, and Space Adventures by revealing a fully modular space vacation station called "Aurora Station" for trips to space that can last up to twelve days. Meanwhile, Virgin Galactic's Virgin spaceship (VSS) Unity, a suborbital, rocket-powered space plane, will be taking flights closer to the

Karman Line (the official border between Earth's atmosphere and outer space), located sixty-two miles above Earth's surface.[83]

Virgin Galactic will supposedly send people to space very soon. Sort of. Probably. Maybe. Virgin's Richard Branson first made this promise more than a decade ago.[84]

Virgin Galactic's rocket-powered Spaceship Two (SS2) space plane won't actually launch from the ground but will hitch a ride up to an elevation of roughly 9.4 miles via a mother ship. At that point, the SS2 will detach and accelerate to a top speed of approximately 2,300 mph in about eight seconds. Tourists will experience weightlessness and check out a view of Earth's curved horizon. The ninety-minute flight will cost $250,000, but that price is expected to decrease.[85]

Next up is a jolly fellow named Rick Tumlinson, the head of Deep Space Industries, an asteroid mining company that aims to be a gas station/building-supply center/air-and-water stop in space. Tumlinson was one of a group that leased the Mir space station from the Russian government for a few months in 1999. Rechristening the station the MirCorp, the group sent up a Jolly Roger pirate flag. No one—not Russia, not NASA, and not the State Department—was amused. But seriously, Tumlinson and partners have pledged their lives and fortunes to "making the human breakout into space" happen in our lifetimes.[86]

And our old frenemies, Russia and China, are already part of the global space community, although China holds its cards close to its vest. As part of China's twelfth Five-Year Plan, a Chinese lunar rover made its first soft landing on the moon in 2019. Among the country's stated goals is to establish a crewed lunar base and then on to Mars.[87]

In two or three decades, humankind might have a couple of small hotels in orbit, with people checking in and out on a regular basis. The sleeping rooms could be arranged in a ring that rotates at a comfortable 3 RPMs, providing three-fourths of the artificial gravity of Earth. In the next two to three decades, the price of space tourism will come down so much that a flight to LEO will be about the same as a vacation to a Caribbean resort.[88]

By the way, if you are elderly or infirm, you might like to retire to a space resort. You wouldn't be subjected to gravity, so you wouldn't need a walker or wheelchair. After a couple of months, you might not want to be earthbound ever again.

NEW NATIONS

Technologically and psychologically, space colonization will be a test for humanity. The first people to stay on Mars for years will be selected for their physical and emotional skills and basic suitability for the challenge.[89] At some point the base camp will become an actual village with people bonding into a society with their own set of idiosyncrasies, prejudices, lexicon, and rules.

Then there are the societal considerations: laws and a government based on democracy or an outer space corporate empire that may base its political, legal, economic, and social structures on a new sovereign state model, the "company nation." A colony may even declare itself a tax haven from the nations of Earth. Penal colony countries might be created in orbit, for a price, as they should be fairly escape-proof.

If we want to create communities outside of Earth, we will need more collaboration from more people, from more disciplines, from more countries than ever before—and that would hopefully include no ancient flag planting and claim filing. Remember that Buzz Aldrin didn't claim the moon for the United States when he planted the Stars and Stripes in the lunar dirt.

THE UH-OH FACTOR

Space pioneers will have to be very careful to avoid any mini-takeoffs of *Aliens* and not bring microorganisms on board during their comings from and goings to Earth. Such organisms might have unknown but nasty effects on unprotected flora and fauna on a planet where someone didn't wash his or her hands after using the space potty.

A recent survey of bacteria on board the ISS revealed five different strains of Enterobacter, a type of bacteria notorious for its ability to survive our strongest antibiotics, ability to thrive in both aerobic (air) and anaerobic (nonoxygenated) environments, extraordinary mobility, and estimated 79 percent potential for infecting humans. At present they are not immediately dangerous. However, experts aren't certain how much of a threat these bacteria pose since we don't fully understand how bacteria operate in space.

Even if contamination isn't outright harmful, it could compromise any research we hope to conduct on foreign bodies. After all, if humans accidentally contaminate some alien environment with microorganisms the first time we visit, how could we ever know whether the life we find later is native to the planet or has evolved from our alien microbes?

We're currently entering a new era of space exploration, and we need new planetary protection policies lest we forget the lessons of our early explorer ancestors on Earth and the diseases that they unwittingly passed around, and exchanged with indigenous peoples.[90]

No one is asking permission to colonize space. So far, astronauts are only subject to the laws of their home countries. For example, by whose laws would someone be tried for the first space theft or murder? How would mother countries respond to rebellious colonies? Could these settlements become independent? How would various settlements on the moon, Mars, or anywhere else interact among each other and with Earth society?

A key advantage of space settlements is the availability of "new land" and resources to develop. Hopefully, colonists will create thriving, expansive civilizations without war or the destruction of a planet or colony's biosphere and without the thought to ripping off others. The asteroids alone can provide enough material to make available new lands hundreds of times greater than what's on Earth. Perhaps these lands will be divided into millions of settlements, creating wealth beyond our wildest imagination and easily supporting trillions of people.

The challenge to these kinds of questions comes from the fact that alienation of offshoot colonies happens, and future space societies may find themselves in need of entirely novel governmental structures. On Earth, colonization by and eventual secession from governing nations have often been a bloody mess. And try to remember George Santayana' s quote: "Those who fail to learn from history are condemned to repeat it."[91]

elp to restore the

6

THE MOST IMPORTANT COLOR IN SUPERCITIES

Getting Green

Macrosized cities of a hundred million or more will be compelled to set aside large or well-placed retreats of green and natural havens for recreational use and mental health for their citizens. Few people actually realize the extent of the need for people to engage with nature—it's going to be a mental health necessity in nonnatural, built environs. Researchers have discovered that "nature" can make you kinder, happier, more creative, more vibrant, better adjusted, and better connected to other people.[1] It will be an essential to a happier, more fulfilling, and healthier life.

FACTS AND FIGURES

- Both looking at and being within green spaces help to restore the mind's ability to focus.
- The experience of nature helps to restore the mind from the mental fatigue of work or studies, contributing to improved work performance and satisfaction.
- Urban nature, when provided as parks and walkways and incorporated into building design, provides calming and inspiring havens and encourages learning, imagination, inquisitiveness, alertness, and social interaction.

- Green spaces provide necessary places and opportunities for physical activity, which improves thinking, learning, memory, and bodily and mental health.
- Outdoor activities can help mitigate symptoms of Alzheimer's, dementia, stress, and depression; they can improve cognitive function.
- Contact with nature helps children to develop reasoning, thinking, emotional, and behavioral connections.
- Symptoms of attention deficit disorder in children can be reduced through activity in green settings, and "green time" can act as an effective supplement to traditional medicinal and behavioral treatments.
- Natural scenes evoke positive emotions and promote recovery from mental fatigue even for people who are in good mental health.[2]

PLANTS AND CLEAN AIR

Access to clean air from forests and plants could be the new status symbol in superpolluted cities. And that access may turn out to be just another way that the rich can afford to distinguish themselves from the poor, who may be forced to choke on secondhand and grimy ghetto air.

People in higher socioeconomic groups with lower rates of physical and mental health problems are more likely to be living in towns and cities with green spaces than are those in poorer circumstances.[3]

METRO RETROFIT

Streets will be refashioned by introducing walking and cycling routes as alternatives to motor vehicles. The preservation of green spaces will also play a vital role in improving storm drainage and air quality, and it will help combat the urban heat island effect.

Currently, in people-packed and paved-over Los Angeles, part of the ruined Los Angeles River has been revitalized with plants and animals. What were once ugly places are now favorites for walking, hiking, kayaking, and picnicking. These places are a natural and social green jewel.[4]

Another model is Portland, Oregon, which has almost three hundred public parks with more than thirty-seven thousand acres of green space. Hundreds of miles of bike lanes make for an appealing, attractive, and mellow way to get around the city.[5]

In Paris, France, Project Oasis is a radical plan to increase the amount of public green space by transforming eight hundred concrete schoolyards across the city into "islands of cool."[6] Meanwhile, in US cities from Baltimore to Seattle, people are rethinking their alleyways and transforming dreary dead-ends into pocket parks as places of cool connectivity and a little green.

Planters, gardens, green roofs, and other features can be incorporated into building design to cool off and dress up a city. The soft rhythmic movements of trees or grass in light and shade or in a light breeze could help create calm, peaceful areas that would aid patients' recovery, improve workers' outlook, and enhance students' productivity.[7] Bright daylight supports circadian (twenty-four-hour bodily) rhythms, enhances mood, promotes neurological health and alertness, and increases the use of natural light, which reduces dependence on electric lighting.[8]

Green space can play an important role in fostering social interactions and promoting a sense of community that is essential in crowded cities. However, parks without obvious beneficial features do the opposite. If a green space is difficult to get to, has poor lighting, or is not clean, it may be seen as unsafe or inaccessible and probably wouldn't boost a visitor's mood. In fact, it may even frighten people away and be viewed as dangerous.[9]

For some people, going to a quiet park is a way to escape their daily routine, while others use nature to challenge themselves and might prefer something strenuous like mountain biking, climbing, or surfing. Still others may find comfort in nature just when interacting with animals or other people.

GREEN STRAIGHT UP

A vertical forest is built straight up into the city sky from buildings, absorbing tons of CO_2 and producing a wealth of oxygen at the same time. The plant life helps reduce smog, dampen noise levels, regulate temperatures inside and outside of buildings, and can become habitats for native wildlife. During the winter, sunlight can easily pass through the bare plant life, and during the summer, the leaves can block harsh rays that would otherwise pour into comfortable living spaces.

Vertical forestry in built environments offers more than just a horizon of trees. It suggests a greener, better view, enhancing a city for building residents. The trend started with green rooftops, but has grown to encompass all kinds of building surfaces.[10]

Greenery not only is beautiful but also provides us with many other services at no charge. One of the most important reasons to preserve greenery is the simple fact that without it we will have less oxygen to breath. It reduces the amount of pollution that enters the soil, water, and air, provides the ground with a natural erosion protection when strong plant roots penetrate deep into soil and help to hold it in place in the event of floods or other natural disasters.

Green space does not just refer to urban parks. Green space is an umbrella term used to describe either maintained or unmaintained environmental areas, which can also include nature reserves and wilderness environments. But, particularly in city situations, green spaces are purposefully designated for their recreational or aesthetic merits.[11]

Growing research suggests that just about any kind of green space—from hiking trails, forests, and coastlines to soccer fields and local parks—can boost the brain's stash of feel-good drugs, like dopamine.[12]

When it comes to seeking green pleasure, the quality of the space may matter more than the quantity. Some research has identified specific types of green spaces—broadleaf woods, parks that feature water, and areas with significant biodiversity, for example—as the most fruitful.[13]

THE GOOD GREEN SPACE IN CITIES

Access to areas such as parks, open spaces, and playgrounds has been associated with better general health, reduced stress, and depression levels. One study found that people who use public open spaces are three times more likely to achieve recommended levels of physical activity and better mental health than those who do not.[14]

Physical activity has been shown to improve the cardiovascular system, mental health, neurocognitive development, and general well-being; and to help maintain healthy weight levels and even prevent some types of cancer and osteoporosis.[15] Providing urban green may encourage people to spend more time outdoors engaged in physical activity, which is especially beneficial for seniors, children, and those who suffer from obesity.[16]

Parks are also a source of positive economic benefits. They enhance property values, increase municipal revenue, and attract homebuyers and workers as well as retirees. They filter pollutants and dust from the air, provide shade and lower temperatures in urban areas, and they even reduce erosion of soil into our waterways.[17]

An evaluation of the largest eighty-five cities in the United States found the health savings from parks was an estimated $3.08 billion. Philadelphia experienced $16 million in annual public cost savings as a result of stormwater management and air pollution reduction, according to a 2008 report by the Trust for Public Land Center for City Park Excellence.[18]

Parks and open spaces made from obsolete structures make compact city living far more attractive. Old rail lines, unused bridges, and abandoned streets can be transformed into greenways and pocket parks. Gardens planted on unused or abandoned lots or rooftops can maximize limited space.

A network of urban parks, natural landscapes, and open spaces that include protected natural lands, ecological reserves, wetlands, and other green areas are critical to providing healthy habitats for humans. They also preserve regional ecosystems and save wildlife and plants in densely built places amid growing cities.

Buildings should be oriented so that heat and sun glare may be reduced, to help the spaces within to stay naturally cool and improve air quality.[19]

KEEPING IT GREEN, NOT YELLOW

Some cities are trying out a lamppost that incorporates a compost bin in its base to collect food waste and is powered by the methane created by the waste as it composts.[20] Another, perhaps more radical, idea is based on the sentiment that public urination isn't going to stop anytime soon. So one urban designer is calling for the placement trash cans in strategic locations with built-in urinals that funnel liquid waste into a tray containing dried garden grass or wood-ash to be eventually turned into nitrogen-rich fertilizer.[21]

WORKING GREEN

In one study of workplaces, employees in windowless spaces introduced twice as many green elements around them as those who had views of natural areas. Workers report that plants make work settings far more attractive, pleasant, personal, and healthy. The presence of greenery improves employee morale, increases worker efficiency, and decreases absenteeism and sick leaves. A lack of nature views or indoor plants was associated with higher levels of tension and anxiety among office workers.[22]

When comparing workforce reactions in windowless offices but with and without plasma TV "windows" showing natural scenes, participants preferred the offices with plasma-display windows. Increased effectiveness was a result even with this artificial connection to the natural world.[23]

For children, natural settings and backgrounds encourage learning, exploration, and building activities; they improve problem-solving skills, the ability to respond to changing situations, and participation in group decision-making. Attention is clearly important for learning, but many kids have trouble paying attention in the classroom, whether it is because of distractions, mental fatigue, or ADHD. Luckily, spending time in nature—taking a walk in a park and even having a view of nature out the window—helps restore kids' attention, allowing them to concentrate and perform better on cognitive tests.[24] Younger children often use outdoor settings as props for make-believe play that enhances emotional growth.[25]

GREEN AND GOOD HEALTH

More than four and a half million children in the United States have been diagnosed with attention deficit disorder (ADD), a condition that affects social, educational, and psychological growth. One study found that children with ADD who played in windowless indoor settings suffered significantly more severe symptoms than those who played in grassy, outdoor spaces.[26]

Immune systems benefit from direct exposure to natural environments or through contacts with green space. It has also been shown that children with the highest exposure to the outdoors during their first year of life were least likely to have recurrent allergies. Living in areas with more trees and plants was shown to be associated with lower asthma rates.[27]

Encounters with nature also provide mental health benefits for the elderly, including those suffering from Alzheimer's, a type of dementia that causes memory impairment, intellectual decline, disorientation, impaired ability to communicate and make decisions, less tolerance for levels of stimulation, and death. Supportive outdoor spaces such as looped pathways, tree groves, landmarks for direction, well-lit paths with handrails, seating areas with privacy, and low-key fragrances and colors are soothing to dementia patients.[28]

Dementia and stroke patients exhibited improved mobility and dexterity, increased confidence, improved social skills, better sleep patterns, improved hormone balance, and decreased agitation and aggressive behavior after being "in nature" for extended periods. Emotional states such as stress, agitation, anger, apathy, and depression are also improved and reduced after time spent with Mother Nature.[29]

Roger S. Ulrich, PhD, director of the Center for Health Systems and Design at Texas A&M University, investigated the effect that views from windows had on patients recovering from abdominal surgery. Patients able to see the natural world healed faster, had fewer complications, and required less pain medication than those who were forced to stare at walls.

These and other findings from Ulrich's theory of supportive design concluded that facilities should incorporate nature views and nature-related art in patients' rooms, aquariums and greenery in waiting areas, fountains and gardens where patients, family, and staff congregate.[30]

The population of the world's prisons is upward of ten million and rising. Many prisoners are not hardened criminals; some are simply awaiting trial, and some are incarcerated for minor offenses. Social scientists claim that just a short walk in natural spaces with plants and trees can be beneficial to this population and could aid in staunching or altering the destructive behavior of convicts.[31]

Gary W. Evans, PhD, a professor of human-environment relations at Cornell University, studies the effect of noise pollution. Evans found that noisy environments have effects that go beyond irritation and hearing damage. In a study of children, he found that kids attending a school with airplanes flying overhead scored 20 percent lower on word recognition tests.

Evans also found that clerical workers exposed to conversation and other mild office noise showed higher stress levels and did not do as well as those with quieter offices. "City planners, architects and others need to pay more attention to this and other research from environmental psychologists," states Evans. And he believes that architecture has profound implications for human health and behavior. Mental health conditions constitute one of the main causes of the overall disease burden in the United States, at a cost of more than $198 billion per year.[32]

OPEN TO THE ELDERLY

Having a connection with nature along with social interaction is important to the elderly, as loneliness can mean higher mortality rates and depression along with decreased alertness. To promote connections with nature, the elderly should have easily accessible spaces due to their more limited mobility. Consequently, having parks and green spaces in close proximity to their neighborhoods or care centers is especially important.[33]

As the world's population ages, there is an increasing need for community environments to support physical activity and social connections for older adults. The proportion of the population age sixty and older is growing rapidly throughout the world. Physical and social environments can positively or negatively influence older adults' physical activity and social connections. Open spaces in urban areas (green spaces) can promote social engagement, physical activity, relaxation, and interaction with nature—all important elements for physical and emotional health—while reducing health costs.[34]

People seem to prefer a random variety of landscapes with scattered trees, plants, and paths. Those who spend time in a park with a greater richness of plant species are seen to score higher on various measures of psychological well-being than those who visit less biodiverse parks.[35]

OBESITY, DIABETES, BABIES, AND STROKES

A systematic review of sixty studies from the United States, Canada, Australia, New Zealand, and European countries on the relationships between green space and obesity indicators found that 68 percent of scientific papers showed that green space is associated with reduced obesity.[36]

An intervention study using community gardening and nutritional education in one southern state found that 17 percent of obese and overweight children had improved their BMI (body mass index) classification by the end of a seven-week program of engaging in outdoor activity.[37]

There is also evidence showing that no residential access to green space is not good for babies; it is associated with low birth weight and may be a contributing factor to long-term adverse effects.[38]

Living close to green space has been associated with a longer lifespan due to reduced risk of ischemic stroke (the most common type, usually caused by a blood clot in the brain) and higher survival rates afterward.[39]

LOOK OUT AND INWARD WHEN OUTDOORS

Of course, when outdoors one must be careful of increased exposure to pesticides, allergenic pollen, infectious agents in soils, animals, and increased risk of injuries. But most potential injuries can be eliminated or minimized through proper design, care, and operation of green spaces, as well as caution by individuals.

Large differences in occurrence of disease are reported when comparing residents of very green and less green settings, even after factoring in socioeconomic status. Healthier individuals tend to move to or stay nearer to greener neighborhoods.

The brain is the only organ that undergoes substantial maturation after birth. Research shows that natural scenes almost always provoke pleasant emotions and promote recovery from all kinds of fatigue. The experience of nature may also provide some relief for those who experience short- and long-term professional burnout.[40]

The constant din and confusion of city life can be mentally exhausting and can actually dull our thinking. Even a small amount of time in a crowded city or on a freeway can cause the brain to suffer memory loss and reduced self-control (as in road rage).[41] Conversely, brief glimpses of natural elements improve brain performance by providing a break from the intense demands of urban life.

When at work, people's attention is focused on critical information or tasks. Exposure to settings that are visually interesting and relaxing aid in recovery of attention spans. Studies at Stanford University show that just by taking a walk refreshes interest and creative thinking.[42]

SPACE, STRESS, AND DEPRESSION

Various levels of stress can lead to depression, schizophrenia, anxiety, exhaustion, fatigue, and even strain on bodily functions. It can also negatively affect people's perceptions of their well-being, including their own mental health. Physical activity has been linked to improvements in mental health and stress. More than one hundred studies have shown that relaxation and stress reduction are significant benefits associated with spending time in green areas.[43]

Ecotherapy, also known as nature therapy or green therapy, is the applied practice of the emergent field of ecopsychology. In a study conducted by Mind, a mental health charity organization, a nature walk reduced symptoms of depression in 71 percent of participants, compared to only 45 percent of those who took a walk through a shopping center.[44] Studies show that exercise can treat mild to moderate depression as effectively as antidepressant medication—but without the side effects, of course. As one example, a recent study done by the Harvard T. H. Chan School of Public Health found that running for fifteen minutes a day or walking for an hour reduces the risk of major depression by 26 percent.[45]

AL FRESCO ATTITUDES

Activities that include wilderness therapy, therapeutic gardening, farming, ecotherapy, nature-based arts and crafts, and animal-assisted aid can improve self-esteem and self-image, self-control, confidence, empowerment, and clearer decision-making.[46]

People don't have to head for the primal woods to enjoy nature's restorative effects; even a view of nature from a window helps. In one study, office workers with a view of nature liked their jobs more, enjoyed better health, and reported a greater ability to recover from "normal psychological wear and tear." Also, those who had walked in a nature preserve performed better than another group of participants on a standard proofreading task and reported more positive emotions and less anger.[47] By the way, blue spaces (rivers, lakes, and coasts) are as important as green—it is not the color that matters but the opportunity to experience it.

Children who live in greener environments have a greater capacity for paying attention, and they're better able to delay gratification and inhibit impulses. Nature's impact on children with a hyperactivity disorder in middle-class settings showed that children exhibited fewer symptoms after spending time in green surroundings than when they pursued activities indoors or in nongreen outdoor areas.

All told, these research findings suggest that individuals' desire for contact with nature is not just the result of a romanticized view but is an important internal process that appears to make people feel and act better.[48]

FEEL-GOOD DRUGS

Green areas allow for exercise, which offers antistress properties through the release of "feel-good" endorphins. Activity and interaction with people also reduces loneliness and isolation for those who engage in any type of exercise, no matter how gentle.[49]

The communal aspect of green and blue spaces may contribute to social closeness, no matter how casual. And for children the impact is equally positive, helping improve exercise, reduce obesity, ease the development of friendships, and encourage independence and contact with people.[50]

For future building projects, it will be important to keep in mind the importance of plants, water, and sky. These are not just luxuries for the middle and upper classes; they benefit the health and well-being of everybody and should be firmly embedded in the plans of developers everywhere.[51]

DESIGN WITH NATURE IN MIND

Psychologist Judith H. Heerwagen, PhD, is already putting the principles of restorative environments into practice in the work she does as a consultant to designers, companies, and others. She's trying to find ways to make people more psychologically comfortable by "naturalizing" interiors, using natural patterns but rendering them in abstract ways. For example, she replaces bold geometrics with abstract natural patterns in floor coverings and uses branch-like forms overhead to make ceilings reminiscent of tree canopies, pointing out, "We didn't evolve in a sea of gray cubicles."[52]

Even though trends suggest that people are moving back to cities, parks give people a chance to get away from urban stress, noise, and built settings and provide places to stretch the arms and legs while enjoying a little "open space." A relatively small investment in urban green spaces can save governments a lot of money down the line in health-care costs, creating good green vibes that will be beneficial in many ways for their citizens.[53]

ODDBALL URBAN PARKS

Planning and management for green areas are not just about cutting the grass, pruning plants, raking leaves, and picking up garbage. The bigger part of management is how to involve the community in the parks. We need to think of parks more as outdoor community centers where we need to invest in practical and positive surroundings and activities for people trying to fulfill their potential. When we improve parks, we're really improving the quality of our lives.

But in increasingly crowded cities, it can be difficult to find room for parks. Fortunately, there are potential places that can provide the same kind of benefits. In recent research, these spaces are more common than we thought.

Some are uncommon and part of the curious and uncommon make-up of our oddball urban places. The modern urban landscape can contain pocket parks from discarded real estate parcels, green walls, and agrihoods. They can be made up of stranded patches of woods, abandoned military bases and airports, stormwater systems, disused rail lines and bridges, forgotten alleys, walkways, and bridges, roofs and facades of buildings, and places where scraps of land are pieced together like an eccentric, multicolored, inner-city crazy quilt.[54]

Rail parks are popping up, inspired by the success of New York City's High Line, a 1.45-mile-long elevated park, greenway, and rail trail created on a former New York Central Railroad spur on the west side of Manhattan. It was saved from demolition by neighborhood residents as a hybrid public space where visitors can experience nature, art, and design, and it breathed life back into a blighted place.[55]

Besides space for physical activity, there can be space allotted for cultural events such as open-air art camps, dance and theater classes, programs for music and poetry, and the more traditional, multipurpose recreational centers and community neighborhood gardens in which to grow produce and flowers.[56]

Privately owned public open spaces (POPOS) may be accessible to all; they may include plazas, rooftop terraces, greenhouses, sun terraces, and just public seating for the enjoyment of people watching.[57] Spaces for natural green will be planned for in our future super-sky-scrapers, where there may be thousands, even millions, of urban dwellers looking for and needing a "green relief" prescription.

The soothing natural backdrops of soft rhythmic movements of trees or grass in a light breeze or the chiaroscuro patterns of light and shade created by clouds are patterns contributing to a calm, stable mental attitude.

Some of the world's most famous cities are as known for their open space as they are for their culture, such as Golden Gate Park in San Francisco, Central Park in New York, Cal Anderson Park in Seattle, Encanto Park in Phoenix, and Boston Common in Massachusetts—all are central attractions in their own right and examples of more to come in future cities.

The way cities and neighborhoods are designed affects whether it's easy for people to walk, cycle, participate in active recreation, use public transport, and interact with neighbors and their community in general. Nowadays, urban planning decisions have a key role to play in contributing to people's overall health; decreasing levels of disease, crime, and dysfunction; and restoring native plants and animals to the "concrete jungle" of municipalities.[58]

INNER-CITY PARK EXPENSE: A COUNTERPOINT

Although parks are inestimable, they can be expensive and unfair to certain sectors of citizens as well. If elaborate projects are undertaken in poorer neighborhoods, they can harm or marginalize vulnerable residents by forcing them out of their homes as rents and property values rise and the neighborhoods morph and gentrify.[59]

It is important to note that disadvantaged population groups often live in neighborhoods with reduced availability of green space. Studies have shown that these areas tend to benefit the most from improved access to urban greenery that may help to reduce inequalities in health, income, minority status, disability, jobs, education, crime, and other counterproductive factors.[60]

If money is scarce to buy land for more parks, then underutilized and abandoned spaces such as railway corridors, vacant lots, power line easements, brownfields, unused roads, rooftops, and even small scraps of land can be made into functional yet affordable spaces that people can use as petite parks.

Have a look around on your next walk; maybe a small plot or vacant lot near you is just the place for a community garden or pocket park. Green is good for your soul and your neighborhood.

7

BLEEDING-EDGE BUILDING MATERIALS

If You Build It, They Will Come

Anyone who built a house just a few years ago would marvel at the materials that are being used, researched, and planned for building structures tomorrow. Many practices in the construction industry have not substantially changed in fifty years, but serious disruption is about to begin.

Hundreds of new products now in the testing phase aim to bring down the cost, improve the quality, increase the strength, and reduce the length of time needed to build new structures. Once these innovations become the construction industry standard, they will substantially increase the number of new housing developments, hopefully making homes more friendly, affordable, adaptable, and economically feasible.

Companies that don't think ahead will be left behind. In the highly competitive construction industry, there is a dramatic need for rapid, agile adoption of new tools, technology, and ways of doing business.

NOT GRANDPA'S PRINTER ANYMORE

Which builder would you pick if one claimed to be able to deliver a 2,500-square-foot house not in three to six months but in twenty-four hours? With a 3D printer, you can create designs or print 3D models of just about anything under the sun, from a gun to a guitar, and in the construction industry, 3D printing can be used to create construction components or even to "print" entire buildings.[1] Construction 3D printing may allow faster and more accurate construction of complex or bespoke items while lowering labor costs and producing less waste.

One Chinese company has constructed in just twenty-four hours a set of ten single-story, 3D-printed homes formed with a cement-based mixture made with construction waste and glass fiber, at a cost of just $5,000 each. [2]

Three-dimensional printers can even produce entire prefab walls with ready-made compartments in which various electronics and appliances may be installed, based on specific customer requests. There's a lot riding on 3D-printing (or stereolithography) construction methods, which will create ways to build homes for a fraction of the current cost and cut traditional construction time by at least 30 percent. [3]

The groundbreaking method takes soil, combines it with just a few additives, and turns it into a building material with a tensile strength three times stronger than industrial clay. Using this material, the poorest countries in the world could build schools, houses, and even hospitals from the ground we walk on. [4]

One printing technique uses a concrete mixture guided by a computer that directs a tube of material to follow the entire outline of a house in one lap, followed by another and another until it is time for the roof. This is a process called continuous contour crafting. [5] Using this process, MIT researchers created a robotic system that manufactured the basic structure of a building in less than fourteen hours. [6] A US construction company 3D-printed a 650-square-foot house on-site in less than twenty-four hours for $10,000. [7] What's more, this process can also be used with metal, called the WAAM (wire arc additive manufacturing) method: a welder with a nozzle fuses layer-by-layer metal rods, kind of like a giant soldering iron. [8]

UP TO CODE

The 3D-printing construction industry is still in its infancy, so don't expect high-rise condos to begin sprouting up like magic beanstalks overnight. Just a cursory review of research on 3D printers indicates that the industry has problems to overcome, like high energy input, limited materials, too much reliance on plastic materials, relative slowness, and dependence on fossil fuel.

This construction method still has some development to undergo before it meets strenuous building codes. And it has to gain acceptance from the construction industry, which will take more time. Meanwhile, while the larger-scale applications are still being developed, some companies are using the technology to create individual pieces and parts of structures.

PREFAB IS FAB

Soon reinforced plastics will make structures lightweight and corrosion-resistant, with walls substantially stronger and longer lasting than most current forms of construction and with built-in solar and water-collecting abilities.

The expense of excess materials and labor costs have traditionally been part of the cost of doing business in construction. Three-dimensional printing and prefabrication allow builders to get material to specifications without waste, cutting construction material costs by as much as 30 to 40 percent and reducing the total human labor it takes to clean up a building site.[9] A Chinese developer, for example, built a fifty-seven-story building in nineteen days through the use of prefabricated building components, preassembled steel framing, an HVAC system (heat and air conditioning), and roofing.[10]

Prefab components are also environmentally friendly, as they reduce material waste and the number of delivery trips to and from the construction site. As more and more building components become prefabricated and standardized, it will become far more practical to insert robots into the construction process. Consider this: robots are already used to produce expensive, intricate machines that demand precision assembly. These same assembly-line robots can and will soon be used to build and print prefab components en masse. And once this becomes the industry standard, construction prices will begin to drop considerably—probably to the dismay of construction unions.

THE RIGHT RECIPE

The two major challenges in perfecting 3D-printed construction are the material and the machine. The material mixture must be formulated to be strong enough and dry quickly enough to support the weight of each subsequent layer without any layer falling over or being crushed.[11] At present the cost of machinery is still too high for smaller companies, which also lack experience, have a limited selection of materials, and are limited in technology. But one day, large buildings will be entirely printed by a single machine with no on-site human input (save for maybe a finger to push a button).

It's easy to predict that product manufacturers of everything from houses to appliances will try harder and harder to compete for market share based on the disclosure of various chemicals of concern and on building techniques.

Pros

- The equipment and materials for 3D building are improving constantly and their costs are falling.[12]
- Regulation, subsidies, and other public policy measures are encouraging the adoption of 3D printing in construction in many parts of the world.
- By reducing the costs associated with nonstandard shapes, 3D printing gives free rein to the imagination of architects, designers, builders, and consumers.[13]
- 3D printing will reduce the construction sector's harmful impact on the environment: a large proportion of the feedstock (building material) will be 50 percent recycled.[14]
- Through the use of conical, hollow, or honeycomb structures, 3D printing increases tensile strength and enhances thermal insulation.[15]
- With machines doing much of the heavy lifting, labor will be reduced, though not eliminated, as a need for people to set up and run the machines decreases.
- Many 3D-printing construction companies claim their process is faster than using traditional cement and procedures for using it.
- The low costs promised could be a major benefit to developers of affordable housing.
- With printers working off a single digital blueprint, there should theoretically be fewer errors.
- Ideally, printers would use only the exact amount of raw materials needed for each project.
- New 3D-printing technology will soon take hours rather than months to complete a building.

Cons

- As is the case with any revolutionary product or process, adoption of 3D printing in construction has been hampered by slow incorporation into building codes and the imposition of arbitrary bureaucratic standards rather than performance-based results.[16]

- Less demand for both labor and traditional materials affects people in many industries.
- Transportation and setup can be tedious and costly.
- Errors on the digital end can cause tremendous setbacks.

THE NEW RIGHT STUFF

Some of the following construction materials might sound like sci-fi, and some still are. Others are in the research and development stage, and still others are in production. Keep in mind when thinking of alternative building solutions that the future comes sooner than you think.[17] Soon or someday they will be on the shelves of stores and in your homes.

Aerographite

Made from hollow carbon tubes, it is strong but bendable, stable at room temperature, and able to conduct electricity. It can be compressed and returned to its normal size while becoming stronger, not weaker. It can also withstand a lot of vibration—great for materials in aviation and satellites.[18]

Airloy

This super insulator is one hundred times lighter than water and combines the strength and durability of conventional plastics with lightweight, superinsulating properties. It can stop and isolate sound, cold, and heat. Airloy is strong, stiff, tough, and among the world's best thermal and acoustic insulators. It can conduct electricity. It can be made from ceramics, polymers, carbon, metals, carbides, or combinations thereof, and is three to ten times lighter than conventional plastics and ceramic.[19]

Aluminum Foam

Made by injecting air into molten aluminum. It is spongy and has a high weight-to-strength ratio. It can be used as strong, lightweight, and durable cladding (covering).[20]

AshCrete

This is a concrete alternative made of fly ash, a by-product of burning coal. Ninety-seven percent of traditional components in concrete can be replaced with this recycled material. And it's hard as hell too.[21]

Bamboo-Reinforced Concrete

It's a mix of bamboo fibers and an organic resin to ensure the bamboo will not degrade or rot. Growing bamboo absorbs large amounts of CO_2, adding to its potential as one sustainable alternative to steel.[22]

Bioplastics

One foe that's going friendly is plastics made from biodegradable material like vegetation, algae, and cornstarch. Bioplastics appear to have great potential, yet there seem to be an almost equal number of drawbacks. Regrettably, many biodegradable plastics may not biodegrade rapidly enough under ambient environmental conditions to replace conventional plastic made from fossil fuels. But bioplastics use 65 percent less energy to make and don't disperse greenhouse gases. Bioplastics are a hot topic in the scientific community and readers need not look far to find many websites dedicated to their advantages and disadvantages.

Bochar

Black façade panels are made from this waste product, produced when trees are burned in kilns for energy. Fifty percent becomes heat energy, and the rest becomes biochar, used in sanitation, farming, water filtration, and many other applications.[23]

Boron

A one-atom chain is a semiconductor. Boron nitride nanotubes are primed to become effective building blocks in nanoengineering projects. If you pull on a nanotube, it starts unfolding; the atoms yield to a monatomic thread. Release it and it folds back. Boron is used for a myriad of purposes from medicine to rocket fuel igniter to possibly a component of recyclable batteries.[24]

Carbon Nanotubes

A thousand times as flexible as human muscle, more bendable than rubber, a better conductor of electricity than copper, and stronger than steel, carbon nanotubes can be used for everything from easily recyclable proton batteries, to golf clubs, bicycles, wind turbine blades, car parts, solar panels, and airplanes, to name just a few functions. These ultrathin materials have the potential to revolutionize modern electronics since you can fit more nanotransistors than those made of traditional silicone onto a computer chip. Consequently, the phone in your pocket will be obsolete, only to be replaced by one that is twenty-five times faster.[25]

Chameleon Paint

This heat-sensitive coating has many applications and is already used for a variety of safety products and energy-efficient materials. Heat-sensitive paint that has the ability to change color with the push of a button is called a chromic material. The color change occurs when electrons within the chemical structure of the paint are manipulated, and the visual change in color perception is immediate.[26]

CO$_2$ Concrete

This novel concept is still in the experimental phase. It captures the emissions from power plants and turns them into 3D-printing material that is two and one-half times stronger than concrete.[27]

Coconut Husk

Sixty billion coconut husks are discarded by the food industry each year. The husks are high in lignin, which can be fully recycled and bound into incredibly strong hardboard, creating less demand for wood and the destruction of forests.[28]

Concrete That Sucks

Topmix Permeable is porous and can absorb four thousand liters of water into the ground in just sixty seconds and funneled into the city's drainage system to be recycled or drained off. It is about the same price as regular concrete.

Cool Bricks

These porous ceramic bricks soak up water; when air passes through them, the water evaporates and creates a flow of cool air.[29]

Data-Storing Glass

A standard-sized disc (smaller than an eyeglass lens) can store around 360 terabytes of data, with an estimated lifespan of up to 13.8 billion years even at temperatures of 374°F. That's as old as the universe, and more than three times the age of the Earth. The discs store information within their interior using tiny physical structures known as "nanogratings." The glass is waterproof and heat resistant, and data on it will survive unless this very hard glass is broken.[30]

Faux Gravel

This is constructed from recycled plastic ground into granules that have been scooped from the sea and mixed with bitumen (a black viscous residue from petroleum distillation). It is used in roads, driveways, and for ground cover.[31]

Feel the Heat

Waste heat can be turned into electricity via a thermoelectric generator (TEG). This is a solid-state device that converts heat flux (temperature differences) directly into electrical energy through a phenomenon called the Seebeck effect.[32]

Ferrock

Based on iron carbonate, it incorporates largely (95 percent) recycled materials to produce concrete-like building material that is even stronger than concrete. This unique material actually absorbs and traps carbon dioxide as part of its drying and hardening process, making it carbon-neutral.[33]

Fire-Retardant Plywood and Black Graphene Radiation-Protecting Paint

Once treated with Burnblock, a new fire retardant that blocks oxygen, this plywood won't burn. While the paint is electrically conductive, it has also has extremely strong radiation-shielding properties, even blocking mobile phone and television signals.[34]

Floats Like a Butterfly but Strong Like Steel

Microlattice, a honeycomb material that has polymer and alumina composites but is 99 percent air, is so light that a slab of it can sit atop a fluffy white dandelion, and a slight breeze can send it floating through air. It is one hundred times lighter than Styrofoam—but as strong as titanium.[35]

Graphene

This one-atom-thick layer of carbon is one million times thinner than a piece of paper—but strong, flexible, a better conductor of electricity than copper, virtually transparent, two hundred times stronger than steel, and bulletproof. Researchers recently created graphene-coated fabrics that can detect and alert a wearer to dangerous gases in the air. Graphene can also help produce clean drinking water by acting as a filter. But it's expensive and hard to produce in scale.[36]

Growing, Self-Healing Bricks

These products, made of bendable concrete, will keep buildings from crumbling. Bendable concrete is very impressive in its ability to resist damage, but even more impressive in that it grows back and repairs itself. Once a crack appears and water enters, the microbes are activated and they multiply, excreting calcium carbonate to plug holes and prevent further damage.

To grow bricks, sand is mixed in alternating layers with a solution of bacteria, urea, and calcium chloride, resulting in chemical reactions that yield a mineral growth between the sand layers that in turn binds them strongly together into a brick. Growing bricks as opposed to firing them in a kiln at thousands of degrees saves a lot of energy and will slash annual carbon emissions by hundreds of millions of tons.[37]

Hempcrete

A concrete-like material is derived from the woody inner fibers of the hemp plant, which are bound with lime to create concrete-like shapes that are strong and also super lightweight. They also dramatically reduce the energy used to make and transport material. Hemp is also a fast-growing, renewable resource.[38]

Lego-Like Prefab for Structures

Just like toy Lego bricks that "snap" together, the architectural kind will do the same with a layer of mortar-like adhesive threaded with rebar for extra structural reinforcement. One side of each brick can be removed for easy access to the inside.[39]

Metamaterial

This is a synthetic composite material with a structure that exhibits properties not usually found in natural materials, for example, a negative refractive index (such as a cloaking material like Harry Potter's invisibility cloak). It could be used in constructing holograms in cameras and electronics.[40]

Microalgae

This could be harvested to make bioplastics or fuel, without waiting a million years for petroleum to mature. Only about six months is needed to make biofuel.[41] It's environmentally friendly, but efforts to make it profitable have not been favorable so far.[42]

Newspaper Wood

This material is made from recycled newspapers when individual sheets are coated with glue and then tightly rolled into logs. This tough material can be treated like most other wood products by cutting, milling, sanding, and finishing with paint or varnish.[43]

Origami Magic

This has nothing to do with making paper swans or fascinating 3D shapes from a flat piece of paper. In the cities of the future buildings might be made cheaper and stronger simply by making them from paper. Objects can morph from a flat piece of paper to a heavy-duty, 3D structure and then back into a collapsed flat surface at the click of a button. Shapes can totally transform to support enormous weights without damage especially when sandwiched between flat pieces of material like plastic. Even with other materials, like metal, builders will be able to utilize the strength and properties of origami construction.[44]

Paper Untearable

Precise polymer nanostrusses (structures as thin as five-billionths of a meter) can be coated on materials like metal or ceramic. The newly created material possesses characteristics including flaw tolerance and shape memory. For an example, it can make paper unwettable, thermally insulating, and untearable.[45]

Pee Power

Urine might now be called "liquid gold" and can be turned into solid "bio-bricks" (calcium carbonate) by mixing it with sand colonized by species of bacteria that produce the enzyme urease. Unlike regular kiln-fired bricks, bio-bricks don't require high heat, and producing them doesn't spew out greenhouse gases. The more time the bacteria are given to work their magic, the stronger the bricks grow, suggesting that several different types of building materials could be created with this method.[46]

Plaited Microbial Cellulose

A form of bioplastic, this is a mixture of bacteria, yeast, and other microorganisms. Research teams have developed a way to manipulate them into layered structures.[47] Practical applications for buildings are still some way off, but there's investigation into many applications from construction uses to scaffolds for tissue engineering, bladder neck suspension, soft tissue replacement, and artificial blood vessels.[48]

Plastic Stress Power

Mechanoluminescence refers to the phenomenon by which a material will light up when it's put under some form of physical stress. The given material has nanowires etched on one side and a coating of indium tin oxide (ITO) on the other. When the strips flail in the wind, the nano-wires slap against the neighboring ITO. This temporary contact allows electrons to leap from one material to the other, creating a current via a phenomenon known as the triboelectric effect. Covering a 3,230-square-foot rooftop, the strips would deliver enough electrical energy (7.11 kW) to power a household.[49]

Receptive Bioconcrete

A microorganism-based concrete, this can host organisms that produce oxygen while absorbing CO_2 and pollution. Bioconcrete will allow the growth of plants that reproduce via spores as opposed to flowers or seeds, such as lichens, mosses, and fungi. Ultimately, bioconcrete will allow plant life to thrive on buildings in a way that is both more sustainable and more efficient than existing green walls.[50]

Reinventing Wood

More architects are returning to wood to build their skyscrapers and ditching the more traditional cement and steel. A Japanese firm unveiled its designs for what could be the tallest wooden building in the world—a 1,148-foot skyscraper—made of 90 percent specially processed wood and 10 percent steel. The building would likely be made of cross-laminated timber (CLT), a material made of many sheets of wood glued and compressed together. The final result is a plank that's more robust than steel, twelve times stronger than natural wood, and ten times tougher than steel or even titanium alloys. It's also comparable to carbon fiber but much less expensive. An outside barrier of plastic functions to protect the wood from rotting. Compared to approximately 63 percent of a tree that can be used in solid lumber, composite panels can allow for more than 95 percent use. And because wood is lighter and easier to transport than steel, it requires fewer fossil fuels to transport it, further reducing emissions.[51]

Replicating Robot Skin

The Terminator is here—or is closer than you think. Scientists have come up with a self-healing skin; a soft, flexible electronic material that can automatically repair itself when damaged. The material relies on photosynthesis—the same biochemical process that plants use to turn sunlight into glucose. It is a thin film in which scientists have embedded chloroplasts (parts of plants that carry out photosynthesis). When ready, it will likely enjoy widespread use in car trims, cell phones, and fabrics, and maybe robots: when their surfaces become cracked or scratched, the film, after exposure to air and sunlight, will easily fill in the gaps.[52]

Saving Buildings with Sound

Fires may be snuffed out soon with deep-toned sound. The new fire extinguisher doesn't blast out any compressed chemicals; it just uses a loudspeaker the size of a subwoofer. When pointed at flames, it looks like it swallows them; actually it is depriving the fire of air. An added bonus is that there's no chemical residue or water.[53]

Seaweed Plastic

Biodegradable plastics made out of seaweed could finally give the oceans pollution relief. Researchers have found a way to create a bio-plastic using seaweed, an accessible, sustainable resource. Their promising new approach may both reduce strain on the plastic-clogged oceans and reduce the earth's dependence on fossil fuels.[54]

Self-Healing Concrete

A type of fungus called *Trichoderma reesei* is being used in a new technique to fill the cracks that develop in concrete and create a self-healing process. When performed successfully, the organism is activated by water and produces calcite, a component of limestone that fills the crack. It's a low-cost, antipollution, and sustainable material. At Purdue University, researchers are adding cellulose nanocrystals derived from wood fiber to concrete. It increases strength, impact resistance, flexibility, and inhibits the invasion of water, a poison to steel.[55]

Self-Healing Paint

It uses chitosan, the main component of the exoskeleton of crustaceans (lobsters, crabs, shrimps, etc.). The chitosan is chemically incorporated into traditional polymer materials, such as the ones used in the clear coatings on cars. When a scratch occurs on the outer coating, the chitosan responds to the UV component of sunlight, filling the scratch.[56]

Self-Healing Plastic

This plastic has microcapsules embedded in it, and they activate when it cracks, releasing resin and a catalyst. These materials fill the crack, getting the original material back to normal in three hours.[57]

Shifting Shape

Researchers have introduced a material that assembles itself into a cube when coming into contact with water. Imagine water pipes preprogrammed to expand or contract in order to change capacity or flow rate, or even to undulate like muscular peristalsis to move water or unclog pipes.[58]

Shrilk

Made from leftover shrimp shells and proteins derived from silk it can form strong, transparent sheets that are biodegradable and even enrich the soil. It may be an environmentally friendly alternative to plastics.[59]

Smarter Windows

Self-powered by thin, transparent solar cells, they control the amount of light and heat entering a building. They can help save on heating and cooling costs.[60]

Solar Film

A second-generation solar cell, or "solar concentrator": by depositing one or more layers of photovoltaic material on a surface such as glass, plastic, or metal, this film can be used in the screens of electronic gadgets and in windows and doors. The stuff takes advantage of nonvisible wavelengths of light—ultraviolet and near infrared—pushing them to the solar cells and resulting in energy.[61]

Solar Paint

SolarLayer is an additive for paints, coatings, and flooring that transforms any surface into a solar energy receptor. Through its application, any roof, wall, street, or path becomes a photovoltaic generator. This technology is designed to work in 3V or 12V systems and has a life expectancy of more than twenty years.[62]

Solar Shingles

Designed to look like and function as conventional roofing materials, these act as a type of solar energy known as building-integrated photo-voltaics.[63]

Stabilized Soil

Made of water-based styrene acrylic polymer, when combined with water it can effectively bind dust and stabilize soil in place, allowing for dirt roads to be made safe, stable, and dustless. Soil stabilization is one of the prime objectives of AggreBind soil and is being used to construct an entire road by utilizing the binding properties of clay soils.[64]

Stanene

A single layer of atoms with tin in place of carbon, it conducts 100 percent of electricity and revolutionizes microchips increasing power with no heat loss of energy. It may replace silicon as a cheap and abundant material for computer chips.[65]

Structural Insulated Panel (SIP)

The major components of SIPs—foam and oriented strand board (OSB)—take less energy and raw materials to produce than other structural building systems. They boost the panel's insulation values as much as 20 percent.[66]

Super Bamboo

It's inexpensive, grows fast, and is surprisingly strong, which is why researchers are looking to get more bang for their bamboo buck when it comes to its uses in construction. The material at the edges of a bamboo rod is actually denser and stronger than the stuff in the middle; it works well as a secondary building material, like plywood, to make houses and buildings stronger, cheaper, with less environmental impact.[67]

Superwaterproof Material

Made up of tiny cones, it not only repels water but also can stand up to extreme changes in temperature, pressure, and humidity. Water droplets bounce off the material, carrying dirt with them, which makes it a good coating for cars, boat hulls, medical devices, and windshields.[68]

Thermal Boost

Thermal bridging (loss of heat due to a break or penetration of building insulation) is one of the primary causes of energy loss in a building. Nanogel-filled polycarbonate sheets (composed of synthetic or biopolymers) increases the insulation factor using less material and less energy than glass. They are 250 times more impact-resistant than glass is and can withstand extreme weather.[69]

Thermal Sponge

Three-dimensional sponge on printed façades optimizes a building's thermal performance in various climate conditions. It works by integrating air cavities for thermal insulation. One manufacturer describes this material as offering "thermal conductivity along with electrical isolation. It has excellent conformability to irregular surfaces and a clean release from most materials. Available in multiple thicknesses to fill various air gap heights. Its cellular structure provides extremely compliant gap filler."[70]

Timbercrete

Made of sawdust and concrete mixed together, it is lighter than concrete and reduces emissions from transportation. Sawdust, a waste product, replaces some of the energy-intensive components of traditional concrete. It can be formed into traditional shapes, such as blocks, bricks, and pavers.[71]

Transparent Aluminum

This substance is three times harder than steel, four times harder than fused silica glass, and 85 percent as hard as sapphire. It is used in bulletproof glass, infrared domes in spaceships, space stations, skyscrapers, and vehicle cockpits.[72]

Transparent Wood

Researchers have found a way to strip the lignin (an abundant and natural constituent of the cell walls of almost all plants on dry land) out of wood and replace it with a synthetic polymer, making it 85 percent transparent. The thermal properties are much better than glass, so large panels of transparent wood would be cheap and useful in constructing large buildings.[73]

Triple-Glazed Windows

These superefficient windows consist of three layers of glass with krypton gas in between and do a better job of stopping heat from leaving buildings than conventional window glass does. Krypton gas is a better but more expensive insulator than traditional argon.[74]

Vantablack

Researchers have announced the creation of a new foil-like material so dark it can hardly be seen by human eyes. Its component tubes are each ten thousand times smaller than a human hair and, when put together, they are so densely packed that the result is one of the darkest substances known: up to 99.96 percent of available light is absorbed into the material. It can increase the absorption of heat in materials used in concentrated solar power technology as well as in military applications, such as thermal camouflage, and in certain construction materials, like house wrap.[75]

VIPs

Vacuum insulation panels make up a porous core material that is encased in an airtight envelope. The air trapped in the core material is evacuated, and the envelope is then heat-sealed. The core material prevents the panels from crumbling when air is removed. Some VIPs are predicted to maintain more than 80 percent of their thermal performance even after thirty years.[76]

Wall Panels: Next-Gen

Panels for wood-framed homes may soon require 40 percent less wood product and may generate 98 percent less waste. A combination of insulation on the exterior and spray polyurethane foam (SPF) in the wall cavity functions as a weather-resistant barrier.[77]

Wave Benders

Researchers have developed a new way to control "elastic waves" that could protect structures from seismic events. A team of scientists worked a geometric microstructure pattern into a material made of steel plate, so that it can bend or refract elastic and acoustic waves away from a target during earthquakes. According to researchers, this technique will save structures from damage caused by earthquakes or tsunamis, which could save lives in residential buildings and other infrastructures.[78]

Wool and Seaweed Bricks

Developed with the goal of obtaining a composite, using abundant local material, that would improve brick strength, wool and a natural polymer found in seaweed was added to the clay of bricks. They become 37 percent stronger than other bricks, more resistant to cold and moisture, and are sustainable and nontoxic. They also dry hard, and since they don't need to be fired like traditional bricks,[79] they save energy.

IDEAS INTO REALITY

Inventors, researchers, and engineers around the world continue to discover alternatives to traditional construction materials and techniques. Though many we have listed are being used, more are still in research, development, and testing phases. These revolutionary technologies and materials will change the way we think about and build structures today and tomorrow.

8

GETTING SOMEWHERE
FROM SOMEPLACE

New Age Transportation

A developed country is not a palace where the poor have cars. It's where the rich use public transportation.
—Attributed to Gustavo Petro

In 2014, Americans spent an average of forty-two hours per year sitting in traffic jams. In Washington, DC, the average was nearly double that, at eighty-two hours per year, and in the West, Seattle now has the second-worst traffic in the United States. Travel plans can be checked on your smartphone, tablet, or computer via Bluetooth; you can tap into your transportation apps for an accurate picture of how to best get around, but traffic can't be improved.

For 95 percent of the time, private vehicles are stationary, parked in lots or in front of houses—or in traffic jams. All in all, the space they take up totals about thirteen thousand square kilometers, which corresponds to an area bigger than Puerto Rico. Los Angeles alone has dedicated 14 percent of the county area to autos. This space could be used for playgrounds or urban gardening—creativity knows no bounds for converting such free spaces.

Public transportation is a necessity, a pillar of green living, and some left-leaners think it should be a right. When asked to identify the top issues of concern with regard to urban environments, 58 percent identified air quality, 53 percent noted traffic congestion, and overall 91 percent claimed that they were concerned about the effect of traffic congestion on their business and their air quality.[1] The largest pollution-producing sector in the country is transportation.[2]

In future cities, moving people will be done effectively, efficiently, and economically and will play a key role in making communities sustainable. For the health and well-being of citizens, walking and cycling between their homes, workplaces, shops, and schools are already being encouraged. Imagine a future where self-driving cars, trains, buses, subways, and EV (electric vehicle) devices are all seamlessly connected through an app, where traffic jams are a thing of the past, and "car parks" have been turned into real parks and green spaces.[3]

SOME CHOICES ON GETTING THERE

One option is called the hyperloop. It's a new technology that can move people at aircraft speeds for the price of a bus ticket. A levitated pod uses an electric motor to glide silently through a low-pressure tube. The hyperloop is designed to operate on demand rather than on a fixed schedule, reducing wait times.[4]

Another is podcars, also known as personal rapid transit or PRT: smaller groups of people are transported in cars powered by electric motors operated by computers and moving on lightweight tracks. Riders select their destination prior to riding, and each car only goes to their final destination without stops on the way.[5]

The Olli is an electric mini powered by IBM's Watson technology. It's operated and paid for through an app. Wherever you are is your bus stop. Olli can also have a conversation with you to provide a dinner recommendation or even tell you to bring an umbrella. Olli had trial runs in Copenhagen, Las Vegas, and Miami in late 2016.[6]

If roads are too congested, go up. Overhead transportation literally takes people off the road and allows them to take advantage of the underutilized urban space above our roadways. Like the subway, it allows for the addition of capacity without displacing existing supply. But it does so with a much more attainable profile of routing, cost, and construction.[7]

Vehicles will become part of an automated smart system called Interactive Driver Assistance, a "connected" system designed to increase situation awareness and monitor and mitigate traffic through vehicle-to-vehicle and vehicle-to-infrastructure communications. In other words, vehicles are going to talk to one another and to services in their surroundings, to smooth the trip of passengers.[8]

Sophisticated scanners will be able to identify you and then charge accounts accordingly. Just think about what this will mean for pedestrian traffic flow when you can walk into a train station without swiping at the turnstiles. We're starting to see technology like this on highways the world over, where cameras are replacing toll collectors, so it's only a matter of time before this translates to public transportation as well. Imagine a city where drones are capable of route mapping, suggesting the best alternatives and means of transportation that lead to less congestion, to get you where you're going.

The following are either in use somewhere in the world or in development:

- Smart stations: transit hubs that bring together a variety of transportation services, from public transportation to self-driving EVs to autocopters, hyperloops, and subway systems to help residents and visitors
- Apps or kiosk data centers that integrate all public and private transportation providers, connecting travelers to the best mobility option for them and providing integrated payment options and real-time travel information
- Traffic information boards and electronic signs displaying real-time information about traffic, parking conditions, issues associated with major events or incidents, and possible detours
- Traffic sensors and signals that can receive and transmit data, and technology installed in fleet vehicles to collect data on traffic conditions
- Sensors on highways and arterials running via a smart system that will monitor traffic not by analog "wait times" but by real-time traffic conditions
- Efforts to make more opportunities for car sharing, scooter using, and bike riding[9]

CARMAGEDDON

Practically all cars in the future, save for specially licensed permits for collectors, will be AI (artificial intelligence) based, self-driving, and autonomous. After all, the US Department of Transportation claims that 94 percent of vehicle collisions are caused by human error, resulting in thousands of deaths every year.[10] And fewer accidents aren't just about safety. In 2015 alone, the incidental costs of motor vehicle accidents equaled more than $44 billion.[11]

There is so much traffic in our lives that we spend 127 extra hours per year sitting in a paved purgatory of our own making.[12] Even though the average personal vehicle sits parked for 95 percent of its life,[13] US drivers spend well over $1 trillion on auto expenses, according to *RMI Outlet* magazine.[14]

A TRANSPORTATION LUXURY FOR LUXEMBOURG

A radical and forward-thinking initiative to make public transportation free for everyone could decrease traffic while helping the environment. Luxembourg is a tiny country, but that doesn't prevent it from having a big-city traffic problem. It also has the highest car-to-person rate of any nation in the European Union. In addition, more than 170,000 people from the neighboring nations of France, Belgium, and Germany commute into Luxembourg for work, swelling the population of its capital city fivefold.[15]

In Luxembourg, the hope is that eliminating all fees for public trains, trams, and buses will encourage more people to use those services. Not only would that decrease personal vehicle traffic but it would also reduce pollution. This initiative won't solve the global emissions problem, but if commuters decide to take advantage of free rides, other nations might see the value in following Luxembourg's lead in addressing traffic woes, the physical damage done by cars, and the woes inflicted on the environment. Luxembourg likes the results, claiming that it has dramatically reduced car use, traffic, and pollution.[16]

PROBLEMATIC PARKING

A staggering 14 percent of Los Angeles County's land is devoted to parking and parking infrastructure, which takes up about two hundred square miles or 18.6 million parking spaces—1.4 times more land than that taken up by streets and freeways.[17] In the United States there are a billion parking spots, four for every car. The downtown areas of most cities devote 50 to 60 percent of their scarce real estate to vehicles.[18]

Parking lots are basically city deserts. They don't employ many people, are a kind of blight on cities, displace neighborhoods, and empty pocketbooks with monthly or hourly parking rates. Some large cities charge as much as $770 a month, and the most expensive rates for parking are in New York—a staggering hourly parking at an average of $27 an hour.[19]

Where parking lots are underused, some cities are already thinking about redesigning and converting them to needed living spaces, which translates into more access for bikes, scooters, and pedestrians and more walking boulevards, green spaces, and additional commercial space.[20]

Autonomous Autos

By 2020 cars with AI will be able to chat with their passengers to enhance the driving experience, will be able to track a user's conversational preferences, emotions, habits, safety, music preference, and may even recommend a place to stop for refreshments.[21]

One thing that everyone working in the auto industry seems to agree upon is that more testing is required, as hiccups have happened with many of the self-driving cars. General Motor's (GM) chief of product development says that by 2020, new cars will be "mostly in charge" of driving and by 2025, "they'll be fully in control."[22] Elon Musk has cut that last prediction by two to three years. Along with his subway innovation, Musk has promised a pod-type car that will be able somehow to traverse tunnels. He believes the best way to eliminate the scourge of traffic congestion is with electric, autonomous vehicles bearing an extra set of wheels, shooting through thin tunnels at speeds up to 150 mph.[23]

Hydro Cars

Hydrogen fuel cells were supposed to be the next big thing in engine propulsion, but the promise peaked during the gas crisis of the 1970s and hydrogen fuel never really gained traction. It was simply too expensive to make the fuel. A couple of car companies have come out with hydrogen-fueled cars as hydrogen-based fuel has become cheaper to manufacture and perhaps promises to become a viable, widespread source of clean energy in the future.[24]

Ammonia has recently surfaced as a source of the molecular hydrogen needed to generate electricity. Now researchers have figured out how to extract the molecular hydrogen and generate power without creating the usual pollutants that come from using ammonia (NH_3).[25]

Nuke Cars

Researchers have been working on nuclear- or plasma-powered cars but none has been successful yet. While small-scale thorium-powered nuclear reactors are theoretically possible, none has been designed that could fit in a car. Thorium is a radioactive chemical element that could, in theory, be used to generate large quantities of low-carbon electricity in future decades, and some theorists claim it can aid in powering an engine that runs off noble gases with no pollutants. Compared to the uranium that powers today's nuclear plants, thorium is more abundant and widely distributed in the earth's crust. But just give it another twenty or so years, and you could be driving a reactor home.[26]

Solar-Powered Rickshaws

A solar-powered buggy by Solar Lab combines pedal power with power produced by rooftop photovoltaic panels. Riders get an easy lift to their destination while being protected by the elements, and the driver gets a break through solar energy.[27]

HIGH-FLYING AUTOS

There has been a lot of high-flying buzz about airborne autos since the first was proposed in 1917, when GM had conversations with "air taxi" companies about using electric vehicle technology to create flying cars—way before George Jetson's flying car.

We're a lot closer to flying cars than you may think. In fact, the country of Dubai has already begun testing a prototype of a self-driving hover-taxi with the hope of launching an aerial shuttle service with an electrically powered quadcopter that can travel on a programmed course at 60 mph at an altitude of three hundred to one thousand feet. The service is meant to help reduce traffic congestion and was built to withstand the country's extreme temperatures during summer.[28]

German carmaker Porsche has a flying passenger drone concept in the works that supposedly won't require the driver to have a pilot's license to operate, and Volkswagen has already been working with Airbus on a Pop-Up car-drone hybrid.[29] Musk's SpaceX claims it will have a working autonomous flying taxi by 2026.[30] Daimler, the parent company for Mercedes-Benz, has teamed up for a vertical takeoff and landing vehicle (VTOL).[31]

In 2016 German authorities issued Volvo's Multicopter a permit to fly. Its first staffed flight took place in 2018 and ended with no problems. The pilot controlled the vehicle easily with a single joystick, and the Multicopter was stable and autonomous enough to retain its position automatically even when the pilot released the joystick. Almost all of the new "flying cars" will be designed for autonomous operation.[32]

The vehicle can reach a speed of up to 62 mph, with eighteen rotors powered by nine independent batteries, and has a one-thousand-pound takeoff weight. The large number of rotors and batteries means that even if one of them fails, the Multicopter can still retain height. It is one of the top candidates in the race to become the world's first air taxi.[33]

Most are VTOL flying cars, meaning no landing strips are necessary; non-VTOLs will need about an 850-foot strip for takeoff and just 160 feet for landing, with speeds from about 300 to 466 mph in the air and a ground speed of around 100 mph with a range of around 450 miles. So it would be possible to fly the 400 miles from Los Angeles to San Francisco in little more than an hour, which is less than a commercial flight and cuts drive time by almost a full workday.[34]

A common denominator for all drone designs is the use of electric motors for quiet, emission-free mobility. This makes them lighter and smaller than helicopters; they are also easier to operate with less complicated technology.

Flying cars will face significant regulatory hurdles. Owners will probably need a pilot license to operate these vehicles and access to a runway in most countries. Cost will vary, but these cars won't come cheap—think Ferrari.[35]

Future aerodromes or skyports will have the capacity for up to one thousand flying taxis. And a single flying copter could be parked on top of a roof, a garage, or a high-rise, conveniently ready when needed. The Uber that drove you home will one day do it by air, when the company inaugurates Uber Elevate, its proposed flying taxi service.[36]

Uber envisions a future in which city dwellers can glide over that horrible, time-wasting gridlock in a quiet, luxurious, clean, cheap, auto-operated, and anonymous aboveground air taxi. Initially they will be operated by pilots, who will eventually be phased out and replaced by an autonomous piloting system. Uber plans to start testing the service in a few years. But new technology is needed—some entity will have to design new electric batteries and figure out the logistics for coordinating thousands of flights per day.[37]

Sky-High Car

A Tesla Roadster will literally hover off the ground using SpaceX thrusters, according to Elon Musk. "I'm not joking," said Musk, who is sometimes known for his outré sense of humor. "We will use a SpaceX cold gas thruster system with ultrahigh-pressure air in a composite overwrapped pressure vessel in place of the 2 rear seats."[38] Wonder what it could do in a quarter mile?

SOLAR HIGHWAYS

New, thin solar panels are designed to be glued into or on top of pavement, and only 215 square feet of solar panels would be needed to generate enough power for all public lighting for a city with a population of five thousand. Autos will be equipped with a solar collector on the bottom that can be reenergized—no plug-in required—as it drives along the pavement, over a solar power exchanger.[39]

The composite material is just 0.28 inches thick, making it possible to adapt to contraction and expansion of a thoroughfare as well as assuring the durability and safety of the tile. The surface also offers good grip for car tires, making roads safer. However, the big drawback is cost—about $4.8 trillion to convert the four million miles of roads in the country. The cost could be partially offset by charging customers for charging, and by the system's modern, more efficient technology.[40]

ELF CARS

Nationwide, transportation by bike in the United States is still less than 2 percent, and twice as many men as women ride bikes. So a crossover might attract more attention. It's called an ELF (electric light fun), a hybrid between a bicycle and a car, which is powered by a 750-watt electric motor; has three wheels, pedals, and a solar panel on top; and can be plugged into a regular home outlet. It has a carrier in the back for a rechargeable battery—so no petropower or foul emissions.

The ELF, which meets the federal definition of a low-speed electronic bicycle, travels at about 15 to 35 mph, and burns 586 of your calories per hour by pedaling. And when a little extra boost up a hill is needed, or to take a break from pedal pumping, the electric motor can be goosed to help move you along. Pedaling does not charge the battery (an option in the future).

Although some people have attached child carriers in back of the seat, the ELF currently seats only one passenger. A standard ELF has a battery that can move it fourteen miles without pedaling and without being recharged, although an upgraded battery can go roughly forty miles. The vehicle works great for commuting, errands, and cruising around town; it can pack 600 pounds of cargo.

It's not car-pricey but quite a bit more than a typical bike. A base model costs close to $10,000—pricey, but consider it gets 1,800 mpg.[41] The ELF may not replace cars or even bicycles anytime soon as a primary means of transportation, but it can provide mobility for seniors or people with disabilities, especially on city and country roads.[42]

E-Bikes

The boom in light electric vehicles (LEVs) in Asia and Europe is now reverberating in the United States, with industry watchers expecting sales of cycles of more than one million by 2020.

LEVs look like high-end mountain bikes, but they're heavier than most bikes. Twist the motorcycle-like grip throttle and the bike surges with a sudden shot of electric power. The top speed of a LEV is restricted to 20 mph, and it offers 11.4 ampere hours of charge. At present the tab is around $5,000.[43]

E-Buses

Electric buses offer more than twenty miles per gallon compared to their petropowered cousins. Against a traditional diesel bus, you're saving more than 80 percent of the energy used, and maintenance is significantly better than with fossil-fuel vehicles.

There are more than twenty transit agencies in US cities that have said that they will not buy more fossil-fuel buses and are looking at 100 percent battery-electric power in their future. New York City got on board with the program and plans to convert its fleet of 5,725 buses to electric over the next two decades.[44]

E-Subways

Inside the city a series of light trains will carry you efficiently and quickly, utilizing underground "electric skate" subway platforms on wheels propelled by multiple electric motors.[45]

Giving a Charge—For Free

Rent or ride a bicycle and then put it back into a kiosk where your leg power is converted into electricity to run the city's electric buses. Such a system would drastically cut down both fossil fuel consumption and greenhouse gas emissions in cities that still use fossil fuels.

DRONES

Drones make most people think of unstaffed aircraft that can fly autonomously—without a human in control. But an unstaffed aerial vehicle (UAV) is a machine that can fly autonomously or by remote control. [46]

Flying drones will share the friendly skies and range from microdrones (smaller than the palm of your hand), minidrones (they can carry small cameras), midsized drones (often used for photography and making videos), large drones (for industrial work and deliveries), and supersized drones (similar in size to a bus or up to a small plane). The designs will be quadcopters (with four propellers), hexacopters (with six propellers), and octocopters (with eight propellers), and the largest ones will be able to move small houses.

By the way, the drone's exact altitude may not seem crucial, but it is unclear if landowners get to decide if they can shoot down a drone over their property at one hundred or even three hundred feet, because no one has actually decided yet who owns the airspace. [47]

Hoverbikes

Another first for Dubai is buying and patrolling its streets with hoverbikes. While they look fun, their range is still very limited, with a flight time of only ten to twenty-five minutes. We have yet to see how such a futuristic mode of transportation could actually be helpful for real-world law enforcement.

The company behind the hoverbikes is California-based startup, Hoversurf. Its FAA-approved Hoverbike is essentially a cross between a motorcycle and quadcopter, with a capped max speed of just over 43.5 mph—and a cost of $150,000. [48]

Marty McFly Fantasy

The use-by date of Marty McFly's hoverboard (in *Back to the Future*) has come and gone—and there's nothing much like it floating around. There are plans available online for a DIY hoverboard powered by a leaf blower that might be improved upon, power-wise, down the road. [49]

A more modern version is not a hoverboard but a propeller-powered drone. Other hoverboards use fields of magnets—a magnetic levitation system. The Hendo, named after its inventor, Greg Henderson, uses the same kind of electromagnetic levitation that allows maglev trains to buzz through the countryside at 300 mph. The main caveat is that, as with all forms of magnetic levitation, you need a special surface for the magnets to push against—you can't just take your maglev train or Hendo hoverboard and levitate over concrete, wood, or water—and you don't want splash down at $10,000 a pop.[50]

There are hoverboards powered by jet fuel or kerosene that is carried on the user's back, but they won't find their way into common use soon, as they are still dangerous and can only fly a very limited duration. So, sorry, no Marty McFlying in sight for some time.[51]

JET PACKS

For all of you James Bond wannabes, first you're going to require the budget of Bond's spy agency because jet packs run from $125,000 to $500,000.[52] One kerosene-powered pack has thrust generated by five miniature jet engines—two mounted off each arm and one affixed to the back, producing almost 300 foot-pounds of force—for two- to three-minute flights, all supposedly manageable and safe.[53]

The latest version of the jet pack, called M3, has miniturbos that fire up three jet thrusters with 1,000 hp output from its thrusters to steer, aided by a sense of balance. Speed data shows up on a heads-up display in the pilot's helmet, and it's steered by adjusting the arm-mounted thrusters. The engines don't burn the pilot because the heat dissipates quickly, and it doesn't take tremendous strength to manipulate the arm thrusters. The inventor claims that anyone can be taught to fly in fifteen minutes. The jet pack is reportedly capable of 200 mph and up to ten thousand feet of altitude in a ten-minute flight.[54]

And it's said to be surprisingly safe; if there is an engine failure it automatically spools down slowly, so the worst thing that can happen is a slow descent to the surface. There is also a built-in life preserver for a water landing—all for a cool half a million.[55]

There are other jet-powered packs, some burning hydrogen peroxide, for example, but capable of only thirty-second flights, and they're not cheap, either, at a reported cost of around $150,000. No matter what anyone says, it's dangerous to strap a rocket to your back and shoot yourself up into the air, knowing you only have thirty seconds of flight time. In addition, the chemical reaction generates superheated steam that shoots out of nozzles at 1,300°F, making it even more risky.[56]

Another version is the rocket belt. It weighs 125 pounds, and the pilot has to weigh 175 pounds or less, or the rockets won't provide enough lift. For a thirty-second ride, it burns seven gallons of hydrogen fuel at $250 a gallon. That's $1,750 for a thirty-second jaunt, still a little high for picking up a six-pack.[57]

So this personal flying product doesn't seem anywhere near ready for public propulsion at a reasonable tab or time frame.

SUPER SPACE PLANES

A space plane is an aerospace vehicle that operates as an aircraft in Earth's atmosphere and as a spacecraft in the vacuum of space. As an air-breathing, hypersonic craft, it could travel at up to Mach 5 (around 3,800 mph), enabling it to cross the Atlantic Ocean in just two hours and the Pacific in three. The plane would cruise at ninety-five thousand feet, with stunning views featuring the earth's curvature at the horizon and the blackness of space above.[58]

China is now testing a hypersonic plane dubbed the "I Plane." Researchers from the Chinese Academy of Sciences in Beijing successfully tested their I Plane (named because it resembles a capital I when viewed from the front) in a wind tunnel at speeds ranging from Mach 5 to Mach 7, or 3,800 to 5,370 mph. In their research, published in the journal *Science China Physics, Mechanics & Astronomy*, the team explains that the hypersonic plane would only need a "couple of hours" to travel from Beijing to New York. A commercial airline flight can take at least fourteen hours.[59]

Global powers are pushing for ever faster flight. With Japan looking to reintroduce supersonic aircraft, several US companies are working on aircraft capable of achieving hypersonic speeds for commercial carriers, and they report impressive progress.[60]

One of the first things that will change in aviation transportation is the speed of a space plane: supersonic at Mach One, or 768 mph; transonic between Mach 0.8 and 1.2, or about 600 to 900 mph; and hypersonic at Mach 5 and above. Strictly available via TV sci-fi is Warp 6, the cruising speed of the starship *Enterprise*, which is 216 times the speed of light.[61]

BIG BUSINESS AIMS AT THE COSMOS

One more billionaire tech magnate is elbowing his way into space. Microsoft cofounder Paul Allen's Stratolaunch Systems (SS) revealed expanded plans to revolutionize the private space industry using a massive space plane.

Stratolaunch, with a wingspan of 385 feet, two airplane bodies siamesed together and powered by six Pratt & Whitney Boeing 747 engines, it resembles a giant catamaran. A spokesperson for SS said it envisions using the plane to carry three rockets and a smaller space plane up to a proper altitude before they detach and fire their engines on a path toward space.[62] For larger payloads, Stratolaunch is planning to have a medium launch vehicle scheduled for 2022 that would be used for satellite launches, carrying up to 7,500 pounds.[63]

Regular commercial service is scheduled to kick off in 2020 with the Northrop Grumman's Pegasus rocket, which has already had more than thirty-five successful launches, but lately it has been plagued with problems.[64]

In a suborbital flight from London to Sydney with the kind of multiple Mach speed being talked about, there wouldn't even be enough time to watch a full-length film. Despite not quite getting into orbit, suborbital passengers would still technically enter space in altitudes of up to sixty-two miles, which marks the start of space that features weightlessness.[65]

It looks as if Virgin is on the cusp of at least suborbital space flight, as it has apparently sold around eight hundred tickets at $250,000 each.[66]

There are several different types of space tourism, including orbital, suborbital, and lunar. So far, orbital space tourism has been carried out only by the Russian Space Agency via Space Adventures, an American company.[67]

The publicized price by Space Adventures (the only firm that has actually blasted "tourist astronauts" to the International Space Station aboard a Russian Soyuz spacecraft) was in the range of $20 to $40 million for a ten-day visit. So far, seven space tourists have made eight flights. Russia halted orbital space tourism in 2010 due to the increase in the International Space Station crew size, with available seats going to actual astronauts.[68]

SpaceX announced that it is planning to send two space tourists on a loop around the moon on its Big Falcon spaceship in 2020.[69]

ROCKET TO YOUR DESTINATION

Elon Musk wants to use his rockets to revolutionize long-distance travel on Earth too. He has suggested that in the future, flying from one place to another on Earth will take from thirty to sixty minutes and the cost per seat will be comparable to what is paid on a conventional jet plane. At about 18,000 mph, with a maximum acceleration of a fairly comfortable 2–3 g's, it would feel like a mild to moderate amusement park ride on ascent and then a smooth, peaceful, and silent descent.

As for space travel, the "xenon ion engine" electrical plasma powered spacecraft that NASA claims to be building is capable of sending a
robot-controlled spacecraft hurtling up at more than 200,000 mph—
and maybe to Mars. The engines function by turning small amounts of
propellant (usually an inert gas, like xenon) into charged plasma with
electrical fields, which is then accelerated very quickly using a magnetic
field. Compared to chemical rockets, they can achieve top speeds using
a tiny fraction of their fuel. [70] To put that into perspective, the space
shuttle is capable of a top speed of around 18,000 mph. According to
NASA, ion propulsion enables spacecraft to travel farther, faster, and
cheaper than any other system. The downside is that it takes about two
years to get up to top speed, which is the time it takes to get to Mars.

TRAINS

Hydro Trains

Elevated hydrogen trains (Hyrail) will be powered by solar energy that
electrolyzes water hydrogen fuel cells. Future cities will have huge
ponds available filled with algae that produce hydrogen, which will run
through stationary fuel cells to provide energy. [71]

Mag-Powered Trains

Maglev is short for "magnetic levitation," which means that these trains
will float over a guideway on a cushion of air, eliminating friction and
using the basic principles of magnetism to replace the old steel wheels
and train tracks. This lack of friction and the trains' aerodynamic design
allow these trains to reach unprecedented ground speeds of more than
310 mph, or twice as fast as Amtrak's fastest commuter train.

The magnets employed are superconducting, which means that
when they are cooled to less than –450°F, they can generate magnetic
fields up to ten times stronger than ordinary electromagnets, enough to
suspend, speed, and safely propel a train. The magnetic fields interact
with simple metallic loops set into the concrete walls of the maglev
guideway. The loops are made of conductive materials, and when a
magnetic field moves past, it creates an electric current that generates
another magnetic field and makes the train horizontally hover about
five inches above the guideway.

Another big benefit is safety. Any two trains traveling the same route cannot catch up and crash into one another because they're all being powered to move at the same speed. Similarly, traditional train derailments that occur because of cornering too quickly can't happen with maglev.[72]

Developers claim that maglev trains will travel from Paris to Rome in just under two hours, reducing train time by almost nine hours.[73]

Bullet Trains

Anyone who has traveled to Japan or Europe knows that bullet trains (named for their shape) can travel at speeds of over 250 mph. In contrast, the United States' Amtrak's showcase Acela train connecting Boston to Washington, DC, slowpokes at just 70 mph. That figure is so low because many sections of the tracks cannot safely support high speeds, even though the train itself is capable of sprints at 150-plus mph.[74]

A Boring Project: The Hyperloop Hype

Super hyper guy Elon Musk aims to completely reinvent public transportation. He claims that he can build a San Francisco–Los Angeles hyperloop for under $6 billion. The trip for San Francisco to Los Angeles theoretically will take less than an hour at a supposed ticket cost of around $20. This time would be almost on par with airlines and cut $80 to $100 off plane fare. However, this doesn't take into consideration legal hassles and right-of-way dogfights that would take as long as building the transcontinental railroad did.

People would enter a "pod," capable of holding about sixteen people, at an aboveground station and travel at speeds from 150 mph to a possible 750 mph. Musk has announced he has verbal permission to build a hyperloop connecting New York, Philadelphia, Baltimore, and Washington.[75]

Musk's Hyperloop One was built on a three-hundred-meter track in a Nevada desert to propel a 1,500-pound aluminum sled in an open-air test. It traveled at 116 mph in 1.1 seconds.[76]

The Rand Corporation's offering has the simple name of the Very High Speed Transit System, an underground tube that would send passengers from New York to Los Angeles in a hard-to-believe twenty-one minutes. There was even mention of a tubelike network that could connect various points of arrival and departure in a vast nationwide transportation network. To do so, it would rely on a maglev system power-propelling cars through vacuum tunnels at speeds of close to 14,000 mph. Skeptics and critics remark that a hyperloop high-speed system might turn out to be an ephemeral and impractical choice for long distance, since the experience of being shot through a tube while pinned to a seat for around thirty minutes without a view might not charm many people. However, cargo does not demand to be charmed and can easily be shoved in a platform and shot through a tube.[77] A hyperloop might not slow down because of weather and earthquakes and might be a less casualty-prone method of cargo transport, with little or no carbon emissions, less construction cost compared to high-speed trains, and its own seductive speed.[78]

The downside is that a democracy doesn't do too well with these kinds of proposals. Each project mentioned here involves a very large infrastructure requiring a lot of people to agree. Doing this kind of project in this country involves tedious and entangled right-of-way and environmental issues. In China or Russia, however, all it takes is someone powerful enough to decide that it is to be done, and it happens.

WATERWAYS

In the future, city traffic could be reduced by taking advantage of nearby bodies of water. One way to ditch traffic could be by employing fleets of self-driving boats that could provide a low-cost alternative to street-based taxis or delivery vehicles.

This idea isn't new, and all the sophisticated info needed for navigation, steering, and docking capabilities can be done inexpensively with low-cost instruments and 3D printers to build docks. The boats could also be built to be interlocking so that, when they're not ferrying people and supplies, they could couple to create rail-type barges, a concert stage, a platform for fireworks, and emergency docking.

ANIMAL BOTS

Like beasts of burden in the past, ani-bots, or machines made to mimic animal behavior, might be the ticket to aid humans in research and exploration and to aid in building and moving supplies on Earth and above. From dragonflies to pack dogs and donkeys to kangaroos, these machines will be human companions and helpmates in the future.[79]

For example, the "Manta Droid," using its pair of flexible pectoral fins, each powered by a single electric motor, is able to swim for up to ten hours. Researchers see it as an alternative solution to traditional, propeller-based thrusters that are used by most autonomous underwater vehicles. It could be an underwater truck, for relaying supplies, or even an underwater Uber, for getting to and from settlements beneath the surface.[80]

And for sheer fun, pretend to be a dolphin—for $500,000. A California-based company has built a "dolphin" enclosure with a rotary engine that can get you up to 20 mph. By diving up and down within a dolphin like capsule, you will feel—and maybe even travel—like the dolphin.[81]

❊ ❊ ❊

One thing is for certain: in the next few decades there is going to be a plethora of choices on how to get where you're going. Let's trust that they will all be fossil-fuel-free.

9

PRIORITIES FOR POWER

Sustainables That Keep On Giving

Energy cannot be created or destroyed; it can only be changed from one form to another.

—Albert Einstein

CARBON KARMA

In the global effort to fight climate change, cities have some of the greatest potential—and the greatest need—to make a difference. Cities consume three-quarters of the world's energy and dispense 70 percent of our greenhouse gas emissions. World experts agree that cities have no choice but to transition into low-carbon or zero net systems if they're going to lower greenhouse gases and slow climate change.[1]

The good news is the International Energy Association believes that world use of renewable energy will grow from 8 percent of total energy use in 2009 to more than 13 percent by 2035. The bad news is coal and natural gas will also grow—energy generation from coal will increase by 25 percent from 2009 to 2035, and shale gas production will grow nearly four times during that same time frame.[2]

The good news is that in the United States, coal use has been declining at about 8 to 9 percent a year.[3] The Environmental Protection Agency (EPA) released the final version of its Affordable Clean Energy rule in June 2019. It's supported by the coal industry, but it is not clear that it will be enough to stop more coal-fired power plants from closing.[4] Another piece of good news is that the use of natural gas has replaced the use of many petroleum products for generation of electricity.[5] The bad news is that fracking, one of the main methods for mining natural gas, is not friendly to the environment, as noted by the EPA.[6]

The additional bad news is that the facts about climate change are being stripped out of EPA information and dialog by Trump. Thousands of web pages with climate change information that were once provided by the US EPA, the Department of the Interior, the Department of Energy, and elsewhere across the government have been removed or buried, according to reports by the *Scientific American*, the watchdog group Environmental Data & Governance Initiative, the *Washington Post*, and *Time* magazine.[7]

THEN AGAIN

It's unlikely, but a Stanford research team believes that we could power the planet entirely with renewable energy by 2050. But we would have to mandate that all new energy production plants use renewable energy by 2030 and convert existing petropowered plants to sustainable energy sources by 2050.[8]

It's quite a wish list to believe that 90 percent of our energy production would come from wind and solar energy and that the other 10 percent would come from hydroelectric, geothermal, wave, and tidal power. Cars, trains, ships, planes, and other forms of transportation would use solar-powered electricity and hydrogen-powered fuel cells.[9]

The only problem is that somehow beating back the fossil fuel industry to a point of significant change is a pipe dream at best. Petro power companies will squeeze the last drop of oil for profits before they abandon it for alternative energy, because no one can claim rights to the power of the sun and the earth. And in 2017, the world subsidized fossil fuels by $5.2 trillion, equal to roughly 6.5 percent of global GDP. That's up half a trillion dollars from 2015, when global subsidies stood at $4.7 trillion, according to an IMF (International Monetary Fund) working paper on fuel subsidies.[10] If governments had only accounted for these subsidies and priced fossil fuels at their "fully efficient levels" in 2015, then worldwide carbon emissions would have been 28 percent lower, and deaths due to toxic air pollution 46 percent lower.

The IMF report suggests a morally grim situation: As the planet careens toward climate catastrophe, governments are forking over trillions of dollars—one-fifteenth of the global economy!—directly to oil, coal, and gas companies. But the challenge of combating climate change through politics is much more difficult than some tidy math can make it seem.[11]

The petroleum industry has enormous resources bound up in fossil fuel infrastructure futures—something like 1.5 trillion barrels of crude and untold trillions more of other shale and tar sand deposits worth hundreds, maybe thousands, of trillions of dollars—not to mention at least one hundred years of natural gas deposits in the United States. Fossil fuel companies are not going to leave that kind of money on the table or under the ground.[12] As far as petroproducts and fossil fuels in the near and far future are concerned, oil company attitudes go something like this: "Solar is great, and we've looked into investing in alternative forms of energy." But until then, every drop of petroleum represents profits. So it pays to delay.

Taking advantage of other alternative energy sources, improving and expanding clean and green transportation systems, and increasing the energy efficiency of buildings are solutions that cities will, however, have to embrace in the future.

Global energy demand is expected to double or even triple by 2050. Hopefully, at the same time, energy demands from unsustainable sources will decrease and a greater source of alternative power will be mined.[13] The scale and rate of this change are major challenges and also present enormous economic, social, and environmental incentives to invent new ways to manage energy and make it more effective and efficient. Some of the following will be quirky, unbelievable, useful, and outright outré, but mostly inevitable.

AN INTEGRATED ENERGY LANDSCAPE

Depending on their locations, climates, existing infrastructures, and available resources, different cities will likely end up using different approaches to tackle their energy needs and to reduce their carbon footprints. When these approaches are used together, the result is a multifunctional physical and socioeconomic landscape of which energy systems are an integrated part.[14]

New York has created a program by which buildings will save energy via sensors, smart meters, and big data analytics (big data refers to managing information from traditional and digital sources on a mass scale to increase productivity and efficiency via automation). One of the recent tests was in the Empire State Building, which was retrofitted with new technology—to the tune of saving almost 40 percent of building energy use and $4.4 million annually.[15]

ALTERNATIVE FUELS FOR THE FUTURE

Some of the following are practical and in use, others are in R&D, and others have been forming in the minds of researchers and scientists looking into the next several decades or into the next millennium for new, clean, and sustainable sources of fuel and energy to power the future.

ATEC

Atmospheric thermal energy conversion (ATEC—the process of converting one type of energy into another form, like solar power to electrical power), will come from built-in windmills; photovoltaic solar cells; self-regulating window shades and solar glass; ground-level waterfalls in buildings for air cooling, humidity control, and power; aerodynamic window systems; and open garden areas on each floor of a building.[16]

Biofuel

Scientists are trying to extract hydrogen from waste materials like vegetable oil or glycerol, the by-product of biodiesel.[17] Replacing gasoline with biofuel from processed garbage could cut global carbon emissions by 80 percent. It's estimated that eighty-three billion liters of cellulosic ethanol can be produced by the available landfill waste in the world. The resulting biofuel can reduce global carbon emissions in the range of 29.2 percent to 86.1 percent for every unit of energy produced.[18]

Feathering Fuel

Every year the poultry industry tosses out 11 billion pounds of chicken waste. Employing that up-to-now unused resource is a process for developing biodiesel fuel from "chicken feather meal"—converting protein and nitrogen and 12 percent fat content into an alternative biofuel.[19]

Body Heat

Eighty percent of body power is given off as excess heat. A resting male can put out between 100 and 120 watts of energy per day, enough to power a Nintendo (14 watts), a cell phone (about 1 watt), and a laptop (45 watts). But so far current technology for converting body heat into electricity is capable of producing only a few milliwatts (thousandths of a watt), which is only enough for small things such as heart rate monitors and watches. When thermoelectric materials can convert low-grade heat into electric energy and charge wearable technology at home, the result will be a reduction in plug-ins at home and lower utility bills.

Sound That Charges

How about a new technology that can take electricity, convert it into sound and send that audio through the air over ultrasound. A receiver attached to a portable electronic device catches the sound and converts it back into electricity.[20] The vibrations created by noise can be converted into electrical energy through the principle of electromagnetic induction. Transmitters could be embedded in all sorts of materials and places, including wallpaper and other household objects.[21]

Bulky Bricks of Power

One surprising, but not novel, way of storing energy is by dropping bricks. When a wind or solar farm makes more energy than the grid needs, an automatic crane on the battery uses the extra electricity to lift a giant brick, weighing 35 metric tons, up to the top of the tower. "When that tower's stacked, that's all potential energy," says Energy Vault CEO Robert Piconi. When the grid needs power, the crane automatically lowers a brick, using the kinetic energy to charge a generator.[22] It's like pumping water and releasing it back down, which creates hydroelectric power.

Squeezing the Sun

In one second, our sun generates all the power that humankind has ever used; it radiates 380 billion billion megawatts of energy per second.[23] By 2015, thirty-seven states had some form of renewable portfolio standard (RPS) goals mandating a percentage of total electric generation from renewable sources. California's RPS is the most aggressive, mandating 50 percent, about 36 percent solar, of total electric generation to come from renewables by 2020.[24]

Solar electricity has reached grid parity (occurring when an alternative energy is less than or equal to the price of power from fossil fuels) in almost half the states in the country.[25]

Getting more work out of the sun via a generator prototype called the beta ray allows twice the yield of a conventional solar panel on a much smaller surface. The design is suitable for inclined surfaces, walls of buildings, and anywhere else with access to the sky. The beta ray can even be used as an electric-car charging station.[26]

Solar Tiles and Paint

The immense surface area of buildings begs for the sheathing of photovoltaics. Solar panel tiles (that absorb solar energy as well as moisture from the surrounding air, splitting the water into hydrogen and oxygen and then collecting hydrogen for use in fuel cells) can cover most parts of a building.[27] Solar paint can convert brick walls into solar energy sources and fuel production for fuel cell autos.[28] Ultrathin film solar cells have now been manufactured that are quite flexible and can fit into or over many materials.[29]

Solar Windows

Transparent electricity-generating veneers can be applied to existing windows. If used in the skyscraper market, which consumes 40 percent of the electricity generated in the United States, this technology could cut energy expenditures by 50 percent and supposedly provide fifty times greater energy than rooftop solar systems.[30]

Photoanodes

Scientists have been working on creating more powerful solar cells by developing low-cost, efficient materials that are similar to the anodes in a battery. That is, they increase the production of solar fuel by aiding the flow of electrons.[31]

The Promise of Perovskites

There has been intense interest in crystals called perovskites that are filled with tiny electric dipoles (a separation of positive and negative charges). When such ferroelectric (having variable electric polarization) materials experience temperature changes, their dipoles (a measure of the separation of positive and negative electrical charges within a system, that is, a measure of the system's overall polarity) "misalign" and cause an electric current. Newly discovered KBNNO, is one such perovskite, which effectively can turn sunlight, heat, and movement into electricity.[32] Someday perovskites could be used to produce solar cells that will have the potential of achieving even higher efficiencies with very low production cost—they're cheaper than silicon—but presently they aren't as physically stable as silicon solar cells[33] and won't be ready for commercial use for several years.[34]

Spacey Solar Sails

A satellite with a 1-kilometer-long wire and a sail 8,400 kilometers wide could generate roughly 1 billion billion gigawatts (10^{27} watts) of power, "which is actually 100 billion times the power humanity currently requires," says researcher Brooks Harrop, a physicist at Washington State University in Pullman.[35]

One gigawatt of power is thought to provide enough energy for about seven hundred thousand homes, and it is estimated that a typical home uses about 11,000 kWh per year. A baseline 1 gigawatt power plant with an uptime of 88 percent (typical for coal plants) will provide 1 GW times 365 days times 24 hours times 0.88 for a total of 7,700 gWh of energy over the year.

Scientists feel that if some of the practical issues are solved, solar wind power will generate amounts of power that no one ever expected. A satellite equipped to tap solar wind power would use a blade attached to a turbine rotated to generate electricity, capturing electrons from the sun at several hundred kilometers per second.

Solar Brick

Guide people to your front door at night with an inviting glow from LED patio pavers powered entirely by the sun—meaning no burnouts or bulb replacement. One day of charge provides each Sun Brick with eight hours of amber-colored illumination, more than enough time to get all of your visitors in and out of the house.[36]

Solar-Powered Trees

Scientists are already proposing that microorganisms that do not exist in nature may someday light and power our cities. Synthetic biology techniques enable the creation of bioluminescence in organisms by manipulating their DNA. Think about creating trees that produce a natural lighting usually found in jellyfish by manipulating the genes of trees.

Not only would we be able to enjoy a mellow light at night in town but also, at home, we'd avoid stubbing toes in the dark and have nightlights for cranky kids and added security. We would also benefit from not having to totally rely on fossil fuels or central power grids to provide lighting for streets or buildings.[37]

Floating Solar

The innovative Hydrelio Floating PV system allows standard photovoltaic panels to be installed on large bodies of water such as reservoirs, lakes, irrigation canals, estuaries, bays, and the seas. This simple and affordable alternative to ground-mounted systems is particularly suitable for communities or industries that cannot afford large land use or don't have commercially viable and consistent sunlight.

The main float is constructed of high-density HDPE thermoplastic (used in the production of plastic bottles, corrosion-resistant piping, and plastic lumber) floats linked together, providing a platform for maintenance and added buoyancy. Microwave lasers would then transmit the energy to a city's grid.[38]

Thermochemical Solar

Thermochemical technology can trap solar energy and store it in the form of heat in chemical molecules. This heat energy can be converted and utilized whenever the need arises.

Researchers are looking into a chemical-electrical process that makes it possible to produce a "rechargeable heat battery" that can repeatedly store and release heat gathered from sunlight or other sources into energy.[39]

Poop Power

There is still energy waiting to be harvested out of our food, even after digestion and exit. Thanks to purple photosynthetic bacteria, we can convert human poop by hijacking a bacterium's ability to turn light into energy—and use it to break down waste into useful fuels. When scientists stimulated the bacteria with a weak electric current, it made the purple microorganisms suck up the hydrogen from fecal matter.

The chemical process extracted the carbon, preventing any greenhouse gas emissions and raising the possibility of new nonpolluting material and an energy source that is currently literally being "dumped."

Extracting hydrogen from organic waste materials, like livestock manure, can be achieved by photofermentation, capturing almost all the methane from the waste products and storing the fuel.[40]

From the Gut of a Gribble

The termites of the sea, crustaceans called gribbles that feed on wood, could hold the key to sustainably transforming cellulose into a liquid biofuel. Research led to the discovery that hemocyanin proteins, which transport oxygen through the bodies of invertebrates, play a major role in the crustacean's ability to extract sugars from wood. When wood was pretreated with hemocyanins, it broke down just as easily as wood that was pretreated thermochemically, a costly process. This discovery may be useful in reducing the amount of energy required for pretreating wood to convert it to biofuel. We could one day convert otherwise unusable wood products (like insect-ravaged forests) into biofuels to power our world.[41]

Nano Methane

With the help of sunlight, water, titanium, and the use of nanotubes as an energy source, surplus CO_2 can be transformed into methane, and when splitting molecules, it releases hydrogen from water (a power source) and creates oxygen (for the atmosphere) as a by-product.[42]

High-Energy Gas

Pyromex Waste-to-Energy technology consists of an ultrahigh temperature gasification process that converts organic content of a waste stream into a high-energy synthetic gas, while the inorganic content is converted to basalt brick for building—a great two-fer.[43]

Hydrogen Power

New technology has provided us with more ways to utilize the abundant resources of water and hydrogen. One method is to produce hydrogen by splitting water atoms, creating a tremendous source of energy that is completely renewable, can be produced on demand, and does not produce any toxic emissions. Its only by-product is water.[44]

Flying Wind Farms

Anyone who sails can tell you that the stronger winds are atop the mast. Swarms of kite-like airborne turbines spinning at high altitudes will send power via nanotube cable tethers back to Earth or to the moon if need be. The stronger winds can generate eight to twenty-seven times the power produced at ground level.[45]

Nukes

Although we think we know how to handle nuclear power plants (that make up about 20 percent of our energy) most us are scared of them. In addition, high costs and bad press remind us of past catastrophes, no matter how rare. Europe runs on nukes, but there have only been two new reactors under construction in the United States in the last several decades while thirty-four reactors have been permanently shut down, and at forty years old, nukes look like a dying breed.

Faux Future for Fusion

This clean source of power holds the promise of eventually producing clean, inexhaustible electricity using a fuel derived from the limitless supply of deuterium from seawater. But decades of research have not solved the problem of fusion not producing as much power as it consumes, and scientists have to continue work on preventing heat from melting the materials that form the fusion chamber.[46]

Some progress has been made. The shedding of heat from inside a fusion plant can be compared to the exhaust system in a car. In a new design, the "exhaust pipe" is much longer and wider than is possible in any earlier fusion plant designs, making it much more effective at shedding the unwanted heat. But at an estimated cost of $25 billion, it seems that fusion has a lot of problems yet to be solved in the near future. According to a World Nuclear Association report, it's more important than ever for the United States and the world to explore practical fusion power—and boost spending on research by $200 million per year—by constructing an experimental reactor and using it to see if the process really does work.[47]

Coffee Power

Every year six hundred billion cups of coffee jump-start the world. The average coffee shop tosses out 22 pounds of coffee grounds a day, the oils of which can actually be used to make biodiesel fuel—to jump-start our vehicles.[48]

Cocoa Power

Chocolate factory waste fed to E. coli bacteria results in the formation of hydrogen, which, as we've seen, may be used for producing clean biodiesel fuel for powering vehicles.[49]

Energy Conversion

When the Sun's energy moves through space, it reaches Earth's atmosphere and finally the surface. This radiant solar energy warms the atmosphere and becomes heat energy. This heat energy is transferred throughout the planet's systems in three ways: radiation, conduction, and convection. It can then be stored, but it cannot be called heat. Thermal energy can be stored by taking a substance, using energy to heat it, and then placing it in a thermally insulated container.

Walking with Energy

Pacing power is produced by flooring that converts the kinetic energy from a footstep into electricity. The energy is stored within batteries and then used to power lighting when needed. Pavegen Systems makes a product that looks like a regular floor tile until you lift the surface and see the hub of circuitry within. Soon you'll able to regularly power nightclub lights by boogying down on floors of kinetic tiles.[50]

Simple Up and Down Power

Elevators will power themselves by collecting energy from their up and down movement and braking. Any excess power can be either stored in proton batteries or put back into the electrical grid of a single building or an entire city.[51]

Large and Tiny Urban Turbines

Solar power and wind power account for the majority of all potential electrical capacity in the country. Skyscrapers provide ideal locations for roof-mounted wind turbines thanks to nearly constant air currents at higher altitudes. A whole new generation of small, ultralight, highly efficient wind turbines can be installed anywhere there is a breath of wind. Industrial giants are racing to build skyscraper-size turbines that can generate ten megawatts or more apiece (enough to power more than one thousand homes). The more powerful the turbine, the cheaper it can generate electricity from a single location. The Haliade-X, a proposed turbine, would stand nearly three times as tall as the Statue of Liberty (or 915 feet) and harness wind with blades that sweep an area the length of two football fields. It would produce enough power for around sixteen thousand households.[52]

Passing Gas for Power

A potential storage technology is the conversion of electrical energy into chemical energy (e.g., in the form of gaseous hydrocarbons), which can be easily stored and distributed in an existing natural gas grid.

Hot Stuff: Molten Salt

Several companies are trying to prove that molten salt can save or generate electricity just as effectively as solar and wind. Facilities that utilize molten salt can operate at any time of day and store energy for up to ten hours. This form of power comes from sunshine concentrated onto a tower by a field of mirrors that heats salt in the tower to over 1,000°F, which can then be used to generate steam and turn a turbine. But it will take some time before it is perfected to save energy reliably and safely.[53]

Water and Wave Power

One of the biggest problems is the difficulty of designing a device to capture the energy of waves, "We may not have even invented the best device yet," said Robert Thresher, a research fellow at the National Renewable Energy Laboratory. Building offshore wind installations, for example, tends to be significantly more expensive than constructing wind farms onshore. Saltwater is a hostile environment for devices, and the waves themselves offer a challenge for energy harvesting as they roll, bob, and converge from all sides in confused seas. No one seems to have settled on a design that is robust, reliable, and efficient.

Despite some steady technical progress scientists say that it might take until 2035 to realize substantial amounts of grid-connected wave power. Hydroelectric (dam) power produces 35 percent of the total renewable and inexpensive electricity in the country. It's readily available, and engineers can control the flow of water through turbines to produce electricity.[54] However, unless you're a beaver, dams have fallen out of favor with environmentalists as damming rivers may destroy or disrupt wildlife and other natural resources. It's doubtful that dam construction proposals, at least in the near future, will overcome environmentalist pressure.[55]

Put Mother Nature's Muscle in Harness

A source of overwhelming power is available but at present impossible to control. It comes via hurricane winds, storm surges, tsunamis, tornados, lightning strikes, the jet stream, volcanic energy, earthquakes, floods, wildfires, avalanches, and landslides, among other catastrophic events.[56]

But if you want to talk dollars, take lightning, for example. In 2009 the world used around 20,279,640,000,000 kWh—more than forty times the electrical energy that all the hypothetically harnessible land strikes contain. So, basically, all the lightning we can capture will give the world enough electricity for only nine days!

To capture each and every lightning strike on land, tall towers (think the nine-hundred-foot Eiffel Tower) around a mile apart in a grid formation covering the entire globe would be needed. That is one tower for each of the almost two hundred million square miles of the earth's surface. The cost for each tower and electrical circuitry storage would total about $82 trillion.[57] Solar is a lot cheaper.

Supercapacitors

Super capacitors can store a large amount of charge and release it at a moment's notice. This makes them extremely useful. Capacitors store static electricity between two insulator plates. Their ability for rapid charge and discharge makes capacitors particularly suited for use in microgrids to stabilize variations in energy sources. However, they can be dangerous when releasing power uncontrolled.[58]

Battle of Boosting Batteries

As gadgets become more sophisticated, there's one technology that seems to be left behind: batteries. Renewable energy sources are likely to become more competitive as storage technology improves and when the sun isn't shining or the wind isn't blowing.

The US Department of Energy has stated that building a battery with the capacity to store energy for less than $100 per kWh would put this source on a par with solar and wind energy.[59]

Engineers are designing a tiny, solid electrolyte battery that allows a charge to flow between two nanoscale electrodes, which could revolutionize portable power supplies and lead to the production of lithium-ion batteries that are smaller than a grain of salt.[60]

Batteries never seem to last quite long enough—and their environmental footprint is toxic. A team from Caltech, NASA's Jet Propulsion Laboratory, and Honda are considering new fluoride-ion batteries to offer a promising new chemistry with up to ten times more energy density than currently available. Apparently, they do not pose a safety risk due to overheating, and obtaining their raw materials has less environmental impact than current extraction processes. But in the best-case scenario, it might take years to bring these batteries to market.[61]

Power to the Proton

A working prototype proton battery combines the best aspects of hydrogen fuel cells and battery-based electrical power. When charging, the carbon in the electrode bonds with protons and electrons generated by splitting water to form a power supply. The protons are released again and pass back through the fuel cell to combine with oxygen to form water to generate power, with no emissions in the process.[62]

The rare earth metals that go into lithium batteries are becoming increasingly scarce and expensive, and mining them can have environmental consequences.[63]

This proton battery can be plugged into a charging port just like any other rechargeable battery. Basically, the carbon footprint left by the battery would be the source of the electricity used to charge it, and the results of mining for the chemicals.[64]

Mini/Microgrids and Energy Distribution

There has always been a kind of social compact with power companies and their customers: pay your bill and, when you hit the switch, the power will be there. The US national power grid is a Frankenstein-like creation, grafted and sutured together on an outdated electrical framework. It is a complex network of independently owned and operated power plants and transmission lines regulated and monitored by the nonprofit North American Electric Reliability Corporation (NERC) and serves more than 334 million people from Canada to Baja California.[65] This system has kept us out of the dark for some time, but it's succumbing to senescence and overload. Meanwhile, a renewing light at the end of the tunnel could cost as much as $2 trillion.[66]

Electrical energy is regulated in terms of demand, the system powering up when one region peaks and lowering when demand is lessened. In addition, the country's copper-wire-based electric grid is basically inefficient. But the grid might not have to be completely rebuilt; it could be replaced by a series of minigrids. For separate buildings, towns, or states, future microgrids may be a way to simultaneously address energy security, affordability, and sustainability through dispersed, locally controlled, independent energy systems tailored precisely to end-user requirements. Each building would operate its own small smart grid that can be connected to larger grids, whether state or national, via the internet of things (IoT), and excess power would be stored in alternative ways for later use.[67]

Working Without a Wire

Wireless power transfer, with little or no loss of power in contrast to copper wire power distribution, has been the brass ring of power transmission since Nicola Tesla tried to introduce it at the turn of the twentieth century. It is technically complicated, but in simplistic form it works like this: There are two coils, a transmitter coil and a receiver coil, involved in wireless power transfer. An alternating current (AC) in the transmitter coil generates a magnetic field that creates a receiving voltage in the receiver coil. Stanford researchers have discovered a practical method through which wireless transmission of electricity may be possible. If the method can be scaled up, that will mean much lower costs in creating and transmitting electricity. Unfortunately, this is a fairly complex process at present and has not been possible on a large scale.[68]

Night Lights

The Chinese plan to launch an "artificial moon" satellite in 2020 that will be eight times as bright as the real thing—with enough brightness to replace all the streetlights in a large city. The artificial moon isn't just some giant light bulb in the sky. A coating on a satellite's adjustable wings will simply reflect sunlight. Both the location and brightness of the human-made moon can be changed or completely shut off if necessary. And since the satellite is mobile, it can assist in disaster relief by beaming light on areas that lost power, potentially saving billions in energy costs and aiding in emergencies.[69]

New Nukes

Bill Gates was working on a pilot project to develop safer and cheaper nuclear power near Beijing that recently came to a screeching halt, thanks to restrictions imposed by President Trump and regulations from the Department of Energy that restrict nuclear partnerships between the United States and China, according to the *Wall Street Journal*.[70]

TerraPower, cofounded by Gates, is trying to build something called a traveling-wave reactor that would be capable of generating fuel out of depleted uranium. While this process would give rise to an era of cheaper, cleaner, and safer nuclear power, the reactor itself would cost about $1 billion to build. "The world needs to be working on lots of solutions to stop climate change," Gates said.[71]

Whatever the choice, there seems to be a bright future for many renewable and friendly energy-source options.

10

PROVISIONING THE POPULACE

Resourcing Resources

First we eat, then we do everything else.

—M. F. K. Fisher

CULTIVATING "CROPS"

According to the World Health Organization (WHO), in the next five decades, we'll need to produce 50 to 70 percent more food. And a lot of it is going to come from where it's needed most—the cities. Well before 2050, farms will be urbanized because that's where the consumers are and they want fresh food as close to "farm to fork" as they can get.[1]

Of all the necessities of a city, the two most important (besides clean air) are food and water. Water can be collected by several means: through a municipal water system; by collecting rainwater; from above and below ground streams; from aquifers, dams, or wells, and by filtering water through a toilet-to-tap system that recycles water, and through desalinization of salt water.

It seems that a conflict between using land for agriculture versus for buildings would ease if the food production achieved via traditional agriculture would be shared by inner-city "agrihoods." There will always be traditional farms, but many are far away from urban areas, which makes it necessary for petropowered vehicles to distribute that produce. And when the "farms" are mega-agribiz companies, it's probable that many use chemicals that are not enviro-friendly to living things.

Traditionally a quarter-acre per resident for farming is needed for a city of just one million; that's 250,000 acres, or 390 square miles, of farmland.[2] One thing that could change this estimate is meat. Animals consume a great deal of food to produce protein, and the number varies by the animal. Assuming we stick mostly with chicken and pork, the need for land might be around 500 square miles for that city. In the case of beef, the conversion rate is really out of whack, based on what cattle eat and drink and the amount of land needed to herd them. It takes 1.5 to 2 acres to feed a cow-and-calf pair for twelve months. It takes 2,500 gallons of water, 12 pounds of grain, 35 pounds of topsoil, and the energy equivalent of one gallon of gasoline to produce one pound of feedlot beef.[3]

In fact, protein from animals is not only expensive but is also one the largest polluters in the nation. One future technology that aims to change this equation is cultured "plant" or "cheat" meat. Other provisions that will find their way onto our plates are alternate edibles made from algae, printed foods (edible ingredients processed from various vegetable bases that are extruded through a nozzle and into shapes), insects, and other exotics that will fill our larders and our bellies.[4]

AGRI-ALTERNATIVES

It is estimated that each calorie of consumed food uses ten calories of petroproduct for cultivating, packaging, and shipping.[5] But there are alternative ways to tap into future farming: there's the family of hydroponics and "vertical farming" on "green roofs" or wall areas of skyscrapers or other high-rise towers. Then there are "brownfields" and abandoned buildings that can serve as outdoor and indoor urban farms and distribution locales.[6]

Tapping into the oceans is going to be critical for food supplies and for our health, but far better husbandry is needed, as the seas are becoming more polluted, overfished, underappreciated, and put at risk. One of the best ways for the continued sustainable use of the open waters is to expand sea and underwater farming—but sustainably.

A by-product of urban cultivation is health benefits for bees, because there will be less need in agriculture for toxic chemicals, one cause of colony collapse that is devastating the bees that pollinate 80 percent of our crops.[7]

Another possible by-product of urban cultivation is the elimination of urban food deserts. Found mostly in poorer neighborhoods, food deserts are where there are few places to buy fresh fruits and veggies and other healthy groceries and where fast food, convenience marts, and liquor stores abound.[8] For starters, education about food choices and what is shoddy for the body should become fundamental in food deserts.

WATER

New products, some made of graphene (or graph hair) make it possible to easily and cheaply make dirty water drinkable in a single pass. Microscopic nanochannels allow water molecules to pass but are too small for pollutants comprised of larger molecules to get by. The filter is easy to use and relatively inexpensive as the material's primary component is renewable soybean oil.[9]

Heightened concern about the impact of future droughts, pollution, and the cost of both is prompting many building designers, owners, and managers to consider ways to further reduce water consumption in the future by using better water-conserving materials. Strategies include improved fixtures, installing rainwater and gray water recovery systems, planting native vegetation in place of lawns or ornamentals, and other innovative approaches to reducing onsite water use and collection.[10]

In any given year from 1982 to 2015, somewhere between nine million and forty-five million Americans got their drinking water from a source that was in violation of the Safe Drinking Water Act, according to a new study. Most at risk: people who live in rural, low-income areas.[11]

WHERE'S THE BEEF?

Americans gotta have their beef; we're second in the world for broiling, boiling, barbequing, and burgers. However, the way conventional meat is produced today challenges resources, the environment, and animal welfare. Global greenhouse gas emissions from the livestock sector increased by 51 percent between 1961 and 2010, spurred by a 54 percent increase in methane and nitrous oxide emissions from livestock manure. Moreover, approximately one gigaton of carbon dioxide equivalent in animal-based foods is expended globally every year.[12] Meat production alone uses about one-third of our planet's freshwater and land.[13]

An alternative may be what is called "cheat meat," "mimetic meat," or "shmeat." Some companies have improved the quality, reduced costs, and potentially reduced greenhouse gas emissions with a new livestock-free process that uses fewer resources to grow pseudo-meats—reducing people's amounts of animal fats ingested and hearts attacked. Some shmeats are plant-based; others are grown from cultured meat cells.

The meat-creation process goes something like this: myosatellite cells, or muscle stem cells, are taken from a cow and put in containers along with fetal calf serum, which is a combo of fetal calf blood and fibrin (a fibrous, nonglobular protein) containing a large number of factors essential for cell growth.

The cells are placed onto gels in a plastic dish, and the calf serum nutrients trigger the cells to split into muscle cells. Those cells eventually merge into muscle and fibers called myotubes and start synthesizing protein. The end product is a tissue strip, described by the *New York Times* as "something like a short pink rice noodle."[14] It doesn't sound too appetizing and, according to those who tried a burger made from tissue strips when they were first produced, it wasn't.

University researchers are working on something called cellular agriculture, the process of using a cell culture to make food products like meat, eggs, fish, and milk instead of using an animal-based process.[15] Our appetites are anticipating the results.

It wasn't long ago that the first test-tube meat could only be had for the billionaire burger price of $325,000 per pound. It took a while, but now it's far more reasonable at about 10 bucks a pound, and hopefully it will be less than half that by 2020.[16]

HIGH-STAKES RESEARCH

It's been five years since the first lab-grown hamburger came into existence, but what no one has been able to do is replicate the texture and structure of specific cuts. The key would be to find a nutrient combination that would encourage the extracted animal cells to grow into a tissue structure comparable to that found in an actual cow. In other words, where's the steak? However, an Israeli company announced that, in a lab, it used cells extracted from a living cow and cooked what looks like a regular beefsteak.[17]

It takes two to three weeks to grow and costs about fifty dollars—but remember the cost of the initial hamburger. One researcher remarked, "The smell was great when we cooked it, exactly the same characteristic flavor of a conventional meat, but a bit chewy."[18]

Among the hurdles still left to overcome is figuring out how to produce test-tube meat at scale. Then there's the biggest hurdle of all:

changing people's minds about pseudomeat. But faux burgers are only a part of the animal-product alternatives that several techno-meat companies are working on. [19]

Researchers are trying to isolate the exact ingredient that gives beef its flavor. It's something called heme (an iron-rich molecule that's abundant in muscle), which just so happens to make up a lot of a burger's meat. Another big challenge was fat, which perks up the taste buds and makes that great sizzle sound. The researchers settled on fattiness from coconut oil and a little beet juice for that bloody residue[20]—a combo of the mouthwatering sniff and stuff that caresses the nose and oozes down the hands.

On average Americans eat about double the amount of protein they need, two-thirds of which come from animal sources. [21] The standard American diet emphasizes a high intake of meat, dairy, fat, sugar, and salt as well as refined, processed junk foods. And we all know what that leads to: diabetes, arthritis, stroke, obesity, and heart disease. So we're not exactly eating for success here. [22] It might be for the best in astronaut health that livestock aren't going to be shot into space.

MOCK MILK

The dairy aisle is changing: nondairy milk sales are growing and cow's milk sales have declined. Milk is really important only for the healthy development of children and, as long as they have the proper nutritional and fat balance, different kinds of milk are fine. [23]

"Mock milk" is produced by the yeast inserted with cow DNA (specifically the DNA that directs its protein-producing properties). The yeast becomes a new microorganism with the ability to produce the same casein and whey proteins (like a cow) when fed the right nutrients. The yeast is genetically modified to contain the genetic makeup of a cow so that it has the ability to produce the same proteins. With the specific DNA that directs protein growth, the yeast follows the same process cows do to produce milk proteins when fed certain nutrients. [24] Supposedly the product tastes just like regular milk and, in terms of protein composition and in theory, it's almost exactly the same as what comes out of cows—except for the moo. [25]

Milk has been derived from an assortment of plants—almonds, soybeans, rice, cashews, peanuts, coconut, oats, hemp, flax, and peas—that are cooked and then ground and sieved to remove the grainy stuff. It depends on each person's taste buds, but nondairy milks and mock milk are healthier than milk fat. [26]

AN ELECTRO-MICROBIOTIC MEAL

A Finnish research team has taken a step toward the future of food by developing a method for producing victuals from electricity. The entire process requires only electricity, water, carbon dioxide, and microbes—sounds yummy. The synthetic food was created as part of the Food from Electricity project that exposes microbes to electrolysis in a bioreactor (a closed system that supports the growth of cells in a culture medium). The process produces a powder that consists of more than 50 percent protein, 25 percent carbohydrates, and the rest water—the texture can also be changed by altering the microbes used in production.

The problem is that, at present, a bioreactor the size of a coffee cup takes around two weeks to produce one gram of the protein. It will be a while before getting up to speed to feed the masses. In the future, it could be used as a means of feeding starving people, and decreasing global emissions, fertilizer, herbicides, and the fuel to deliver them.[27]

GROWING UP AND UP

One solution for farming food is called agri-tecture or vegi-tecture and is based on inner-city farms that grow produce upward and inward. Vertical agrihoods essentially involve planting in stacks instead of on a horizontal field. This method saves space and eliminates the need to acquire huge parcels of land to grow crops. It can also be done inside warehouses and in other indoor as well as outdoor spaces lying fallow in cities.[28]

As far as energy goes, a thirty-story vertical farm needs 26 million kWh of electricity but can regenerate more than 56 million kWh through solar energy, windmills, and a biogas digester that turns food refuse into energy as well as into compost that enriches soil.[29]

The best method for growing, especially in cities, would entail the methods of hydroponics, aeroponics, and aquaponics, which are called soilless plant propagation. All operations are generally confined to the indoors, where environmental factors of water, temperature, humidity, light, and insects can all be closely controlled. It uses no soil, requires 90 to 98 percent less water, produces far better year-round yields, and uses no commercial-farming chemicals.[30] Today's largest vertical farm is located in Michigan on about eight acres and is home to seventeen million plants.[31]

THE "PONICS" FAMILY

Hydro

In hydroponics, nutrients are fed to plants via water in sluice boxes exposed to light. It uses as much as ten times less water than traditional field crop-watering methods because water in a hydroponic system is captured and reused rather than allowed to run off and drain to the environment. Hydroponic plants grow 25 to 30 percent faster than traditionally grown plants because the perfect blend of nutrients is delivered directly to the root system. The plant does not need to expend energy on an extensive root system to find the food it needs, so all of its energy goes into upward leaf growth.[32] Plants can be placed closer together, which reduces the space needed to grow the same amount of crop.[33] The single most compelling reason for gardeners to switch to soilless gardening is its ability to significantly increase crop yield, from two to five times.[34]

A significant amount of fresh produce could be planted and grown in cities (or in confined space and/or on other planets)—with potentially lower costs. For example, an indoor farm in San Francisco can produce 2 million pounds of lettuce each year within a space that's no bigger than an acre.[35] Hydroponics is also a popular system for small-scale home gardens; systems come in kits and are very easy to put together.

Aero

In aeroponics seeds are "planted" in pieces of foam stuffed into tiny pots, which are exposed to light on one end and nutrient mist on the other. The foam also holds the stem and root mass in place as the plants grow. Eliminating soil frees plants and roots, allowing extra oxygen to penetrate and resulting in faster growth. In addition, aeroponic systems are extremely water-efficient.[36]

Aqua

Aquaponics is a symbiotic integrated system that combines cultivating plants in a nutrient aquarium-like environment with aquatic animals such as snails, fish, crayfish or prawns, and ducks that feed on algae and leafy crops. Fish and fowl waste is turned into compost that acts as food for the plants that filter the water for the fish. So the end result is produce, marine food, and even eggs and meat from waterfowl. All of the leftovers, including dropped food and other waste, become potential plant nutrients in the water. [37]

ONE HUNDRED THOUSAND CALORIES A DAY

It has been suggested that a thirty-story vertical farm could feed fifty thousand people, providing two thousand calories for every person each day. [38] (One piece of advice about calories: the factoid about people needing two thousand calories a day is just a guesstimate. Depending on gender, work, body style, age, and genetics, caloric need can vary greatly. [39]) Consequently it will be beneficial for buildings in the future to have space reserved for crops.

Sunlight-reflecting and collecting devices, such as artificial light shelves, light pipes, and fiber optics, can deliver natural light deep underground or to inner-city high-rises to provide energy for both photosynthesis and energy. [40]

DIP INTO THE OCEANS

"We're going to have to change the fact that pigs in America eat more fish than sharks, and domestic cats eat more fish than all the seals in the North Atlantic Ocean," says Paul Watson, of the Sea Shepherd Conservation Society. [41]

With the world's populations bulging, the oceans and their bounty of fish are on the hook for growing global hunger. One answer is aquaculture, otherwise known as fish farming, which has been on the rise over the past few decades to meet the soaring demand for seafood. [42]

Unhappily, the most common type of aquaculture is mariculture, or the cultivation of marine organisms in the ocean or within an enclosed section of water modeled on land-based factory livestock farms. [43] Some of these operations are infamous for their low-quality, tasteless, subpar fish pumped full of antibiotics, polluting local waterways, and fouling the gene pool of other wild fish stock. According to a *New York Times* editorial cited in *Atlantic*, aquaculture "has repeated too many of the mistakes of industrial farming—including the shrinking of genetic diversity, a disregard for conservation, and the global spread of intensive farming methods before their consequences are completely understood." [44]

Providing fish with a continuous supply of clean water and healthy food while reducing the spread of pathogens, contaminants, and toxins allows fish to grow faster, more efficiently, and with far less disease.[45]

Flatfish and shellfish have been farmed for years, and seaweed was a staple food for early American settlers.[46] And by the way, seaweed farms have the capacity to grow massive amounts of nutrient-rich food while using photosynthesis to pull enormous amounts of carbon dioxide from the atmosphere. Some varieties are capable of absorbing five times more carbon dioxide than land-based plants, and they also filter out nitrogen (three hundred times more potent in trapping heat than is carbon dioxide in greenhouse gases). And seaweed is one of the fastest growing plants in the world.[47]

About 50 percent of seaweed's weight is oil, which can be used to make sustainable and clean biodiesel for cars, trucks, and airplanes. The Department of Energy estimates that seaweed biofuel can yield up to thirty times more energy per acre than land crops, such as soybeans, and produce 70 to 80 percent fewer greenhouse gases than natural gas does.[48]

Too many biofuels are produced from food crops such as corn and sugar, which drives up global prices in a world where a billion people are already hungry. Because they require no fresh water, no deforestation, and no fertilizer—all significant downsides to land-based farming—these ocean farms promise to be more sustainable than even the most environmentally sensitive traditional farms. Algae has the potential to produce ten thousand gallons of biofuel per acre.

Scientists at the University of Indiana recently figured out how to turn seaweed into biodiesel four times faster than other biofuels, and researchers at the Georgia Institute of Technology have discovered a way to use alginate extracted from kelp to ramp up the storage power of lithium-ion batteries by a factor of ten.[49]

More About Sustainable Seaweed

Professor Ronald Osinga at Wageningen University, in the Netherlands, has calculated that a network of "sea-vegetable" farms measuring 180,000 square kilometers—roughly the size of Washington State—could provide enough protein for the entire world.[50]

And there are other upsides to seaweed: it's rich in vitamins, minerals, and omega-3 oils, and requires no fresh water. It doesn't cause deforestation, use fertilizer, or create greenhouse gases—significantly unlike land-based farming.[51] If done right, a new generation of green aquaculture is poised to figure among the most sustainable forms of farming on the planet.

Through "oyster-tecture," one oyster can filter thirty to fifty gallons of water a day, reducing total nitrogen in the seas by up to 20 percent, and a three-acre oyster farm filters out the equivalent of nitrogen produced by thirty-five people.[52] Floating gardens and oyster reefs help protect coastal communities from hurricanes, sea level rise, and storm surges.[53]

Sea Pod Beds

Another agricultural alternative that experts are developing is growing plants in pods on the seabed for future food security. An Italian company uses a version of hydroponics while creating freshwater through desalination. As water evaporates, drops condense on the roof of the pod and then drip back down as freshwater to feed herbs and vegetables.[54]

Submersion of plants in seawater pods offers a stable temperature while avoiding exposure to extreme weather conditions on land, pests or insects, disease spores, or foul seepage. And tests carried out by the Ocean Reef Group (ORG) suggest that crops underwater grow faster than their land-based counterparts.[55]

It will take a few years to see if this method is economically feasible and to be careful of some of the vast wild spaces on Earth—the oceans. As we develop our ocean farms for future populations, we will have to be mindful of protecting the wild seas lest they become another exhausted and further polluted resource.

Yum Yum, Pond Scum

The blue-green, thick, jellylike stuff you see in fish tanks, seas, lakes, rivers, ponds, and on and under rocks is not what you would see on your plate. Proponents of the edible algae that is known as spirulina claim it could help provide a sustainable source of protein; is packed with antioxidants, minerals, vitamins, and other nutrients; and is a form of homeopathic medicine, although it can cause a gas attack.[56]

Spirulina is finding use as a food supplement—like in smoothies—and as an alternative to "cheat meat." It grows within a week, whereas it takes six months to grow a kilogram of beef.[57] And ounce per ounce it contains more protein than beef.[58] It is eaten fresh, dried, or cooked like spinach and has virtually no taste, so you can mix it with anything—

kind of like sea tofu. An added bonus is that it also feeds on carbon dioxide.

Algae also show great promise in the area of biofuel. Some ten million acres, about 1 percent of the total amount of worldwide acreage for grazing and farming, would be sufficient to grow enough algae to convert into what would equate to the total amount of diesel fuel in use in the country today.[59]

Floating Farms

Singapore, one of the most densely populated countries with little farmland, has created a futuristic "floating vertical farm." The whole system has a footprint of only about sixty square feet, or the size of an average bathroom. A total of 120 farming towers have been erected in Kranji, fourteen miles from Singapore's central business district, with plans for three hundred more, which would allow the farm to produce two tons of vegetables per day. The system will use floats in local harbors that will provide year-round crops. Its loop shape will enable the vertical structure to receive more sunlight without having significant shadows. The farms will contain a number of sensors that can monitor the crops, sending information on their status to the people or smart systems in charge of tending them.[60]

Fishy-Tasting Fish

You've definitely heard of veggie burgers and tofu dogs, but you may not have heard of fake fish. A *Wall Street Journal* article looked at the rise of imitation tuna, crab, shrimp, and even sham smoked salmon. The surprise conclusion: it's rather tasty. Cutting-edge food science is being used to manipulate ingestible material ranging from tomatoes to pea protein to catch the elusive textures and flavors of fish.[61]

GENE SPLICING AND GMO GOSSIP

A lot of concerned people call it "Frankenfood" and, generally, genetically engineered anything has caused a firestorm of protest. A survey by the Pew Research Center shows that nine out of ten scientists from the American Association for the Advancement of Science say that genetically modified organisms (GMOs) are "generally safe" to eat. At the same time, almost half of all US adults think, Nope, we won't eat them. [62]

We have been genetically modifying our foods in one way or another for thousands of years. Farmers, seed companies, and botanists have improved the genes of plants by saving the best seeds and splicing plants to improve their growing and yielding capacity. For instance, the skimpy kernels of corn of a hundred years ago have morphed into the big ears of juicy corn on the cob today.

We can alter the DNA of seeds and manipulate genes, which result in plants that reject toxic herbicides as well as insects, use less water, and are becoming more robust. Some GMOs are specially made to be packed with extra vitamins, minerals, and other health benefits. Some biotech companies are doing experiments to make meat better for us, such as boosting the number of omega-3 fatty acids in it while decreasing animal fats to aid in preventing heart disease and stroke and attempting to protect against cancer and other medical conditions.

A group of scientists from the National Academies of Science did an extensive review of research on the safety of crops from GMOs over the past ten years. [63] They found no significant harm directly tied to genetic engineering. Another study claimed that GMO corn varieties have increased crop yields worldwide from 5.6 to 24.5 percent when compared to non-GMO varieties. They also found that GMO corn crops had significantly fewer (up to 36.5 percent less, depending on the species) mycotoxins, toxic chemical by-products of crops, which are linked to illnesses. [64]

The American Medical Association thinks genetically modified foods are fine. Part of an official statement notes that in almost twenty years, no clear impacts on human health have been reported or confirmed in professional journals. [65] The World Health Organization agrees. WHO, along with the UN's Food and Agriculture Organization, maintains a set of science-based standards, guidelines, and practices called the Codex Alimentarius to promote good, safe food for everyone. [66]

The Food and Drug Administration takes a slightly different approach to genetically engineered animal products. It has issued guidance to help developers meet the standards of the Codex Alimentarius and US Food Safety regulations. [67] The Center for Veterinary Medicine makes sure that any given animal food is safe to eat. [68]

Although there is not sufficient research to confirm that GMOs are a contributing factor to disease, doctors' groups, such as the American Academy of Environmental Medicine (AAEM), tell us to hold off on making them part of our diets. One claim is that "several animal studies indicate serious health risks associated with genetically modified food, including infertility, immune problems, accelerated aging, faulty insulin regulation, and changes in major organs and the gastrointestinal system."[69]

The American Public Health Association and American Nurses Association are among many medical groups that condemn the use of bovine growth hormone, because the milk from treated cows has more of the hormone IGF-1 (insulin-like growth factor) that has supposedly been linked to cancer).[70] Others fear the contamination of the gene pools of plants. At the same time, there are those who believe that discontinuing the availability of GMOs may be depriving deprived people of the future, putting them face-to-face with famine and malnutrition.

On the other hand, between 1996 and 2008, US farmers sprayed at least 383 million pounds of herbicide on produce, that is, non-GMOs. One of the most overused is Monsanto's Roundup, which has been reported to result in "superweeds" that are resistant to the herbicide. Recently, courts have held that it should be considered a carcinogen.

Recent studies and court verdicts have shown that consumers and workers across the United States have been put at great risk of developing non-Hodgkin's lymphoma (NHL) due to exposure to glyphosate, a human carcinogen used in Monsanto's Roundup weed killer products.[71] And several court cases have agreed, resulting in massive legal compensations.[72]

MORE THAN NAPKINS

Another contribution of genetic modification is edible cotton. For the first time, the cotton napkins on your table will be a form of food. The US Department of Agriculture's Animal and Plant Health Inspection Service lifted its regulations on genetically modified cotton, meaning anyone can now grow its seeds as an edible.

In addition to producing fiber for food and fabrics, cotton also generates 1.6 pounds of seeds for every pound of cotton. This could even eventually pave the way for cotton seeds to be sold at your favorite supermarket, as these seeds contain a ton of protein. Unfortunately, they also contain a ton of gossypol, a chemical compound that protects the plant from pests and diseases but is also toxic to people. But scientists have found a way to genetically modify the plant to turn off the gene that produces the toxin. The seeds, which taste like chickpeas, could ensure that about 575 million people worldwide could meet their daily protein requirements.[73]

PRINTER TO PLATE

There's a great scene in the sci-fi pic *The Fifth Element* when the female protagonist pops a pill on a plate and puts it into a microwave-looking machine and out pops a fully roasted chicken. And that may be possible in the future.

Food 3D printing is the latest trend in a new industry that enables foods like chocolate, pasta, pizza—and just about anything else—to be "printed" in your home on a machine. Food 3D printers are modified with special pressurized tanks to extrude raw material in the form of liquid or paste that can reproduce and customize foods in shape, texture, taste, and form.[74]

This process can be healthy and good for you and the environment because it converts alternative ingredients, such as proteins from algae, beet leaves, or insects, into tasty products. Food printers can make both savory and sweet foods; users can choose from a library of shapes or create their own to print. Up to five food capsules can be loaded into the printer at one time, and you can make a pizza (one capsule) in five minutes.[75]

Whether you think that it's rad sci-fi or the ultimate laziness, just pressing a button to print your lasagna is pretty far out. And these capsules will probably be free of preservatives, with a shelf life limited to five days. Currently, these devices only print food that must be cooked as usual. But future models will also produce ready-to-eat.[76]

They have a touch screen that connects to a recipe site, but users can control the device remotely using a smartphone. So you can prepare food while on your way home or watching TV.[77] The food printers will vary in size, depending on configuration, and one that is the size of a medium microwave will start at about $1,000.[78]

But there are always naysayers. In the '70s, people were a bit fearful of microwaves and thought their food could be poisoned with radiation or something. The same goes for the food printer. This is real food, with real, fresh ingredients, but it's just prepared using a new technology. And according to the manufacturers, "everybody that tested it liked the food."[79]

CRICKET CASSEROLE

Because of the growing need for quick and cheap sources of protein in the future, many people (and their pets) may have to redirect part of their diets to the world of insects, the largest part of the animal kingdom. And before you say, "Ick," consider that the eating habits of nearly one-third of the human population includes insects as part of the daily diet.[80] They provide protein and omega-3 fatty acids that are more than comparable to the amounts found in meat and fish.[81]

Insect farming is a very inexpensive, efficient, and sustainable way to produce food. Most insects can be raised using waste from slaughter-houses, grain mills, food processing plants, and restaurants. It takes 20 pounds of grain and 2,500 gallons of water to produce one pound of beef, 10 pounds of feed and 576 gallons of water to produce one pound of pork, 5 pounds of feed and 468 gallons of water to produce one pound of chicken, and 3.73 pounds of feed for catfish to reach market size at 1.5 pounds.[82] We need to find alternatives soon; already all seventeen of the world's major fishing areas have reached or exceeded their natural limits.[83]

Crickets, on the other hand, require only one-half pound of food to produce one pound of body weight. Also, 80 percent of a cricket's body is edible, compared to only 55 percent of the bodies of chickens and pigs, and to 40 percent for cattle.[84]

Thirty percent of the world's land mass is presently used to graze or raise food for livestock. Insect farming requires far less area and can be completed in small buildings with controlled environments. Insects emit far fewer greenhouse gases and ammonia than livestock, making insect farms much more environmentally friendly.[85] Worldwide there are an estimated 2,100 species of insects that are considered edible.[86] This allows for production in urban industrial sites and is also especially attractive for those living in the tight confines of alien spaces, like colonies on other planets.

Zoonotic diseases (ones that normally exist in animals, but can infect humans) have caused widespread epidemics in many parts of the world. The potential for zoonotics is unlikely with insect farming. First of all, insects are more distantly related to humans than mammals are, and they are cold-blooded, which makes the adaptation of zoonotics from insects difficult. Diseases can be transferred by insects. But, for the most part, when it happens, it can be eradicated with far less trouble.[87]

Cricket casserole or chocolate-covered ants, anyone?

11

SMART CITY SYSTEMS
AI and Automation

The best way to predict the future is to design it.
—Buckminster Fuller

IN SMART SYNC

Watch out. The walls have ears, the house has eyes, and artificial intelligence (AI) can bust you. "Smart house" is the term used to define a home whose appliances, from lighting to security systems, are capable of communicating with one another and to you and can be controlled remotely by a programmed schedule from an electronic device like a computer, tablet, smartphone, or even a watch. All devices are controlled by a master home automation controller, often called a smart-home hub.[1]

For example, the fridge will tell when you're out of orange juice or eggs, put the item on a list, or go ahead and order groceries that, once a market delivers them to your home, perhaps will be put away by a house-bot.[2] Walls will do far more than hold up the roof and provide a border with the outside. Doors and floors will be "track padded" with sensors that will recognize footsteps and track where you are going and open doors. Through "airborne electromagnetic noise," walls can track a user's gestures and touch, and record conversations.[3] And if you choose, the house will be surveilled with strategic minicameras.[4]

HISTORY OF SMART HOMES

The first smart device debuted in 1975 and was called X10; and worked with programmable outlets and switches and controlled some appliances, lights, and temperature.[5] A 1999 Disney film, *Smart House*, provided mainstream audiences with a sense of the possibilities on the horizon,[6] but the first smart-home models began to hit the consumer market in the early 2000s and have proliferated via the internet and related technologies today.[7]

SMART HOMES TOMORROW

As people travel for business or pleasure, chauffeur children to their school and activities, and pursue their own social pastimes, new smart systems will provide connectivity to everything in the household and will take care of household tasks. By glancing at their phones or other electronic devices, homeowners will obtain a quick check of the premises.

Any device in the house that uses electricity can be connected to a smart-home network. Whether you give a command by voice, remote control, tablet, or smartphone text, the smart system reacts. Most applications relate to lighting, home security, home theater and entertainment, and thermostat regulation. Add cameras and acoustic, aural, or olfactory sensors attached to AI programming, and you have a house with which you can buddy up and schedule when dinner or drinks will be ready, make and confirm appointments and other future plans, handle kids and their needs, ask advice, and even get a heads-up on your health.[8]

The idea of a smart home might make you think of George Jetson or maybe Bill Gates, who spent more than $100 million building his smart home.[9] But the days when smart-home systems were only for the well-to-do, tech-savvy, or the wealthy are past. Smart homes and home automation are becoming more common and will be more so in the future.

Much of this is due to the smashing success of smartphones and tablet computers connected to the internet of things (IoT)—the network of devices, vehicles, and home appliances that contain software, actuators, and connectivity that allow these things to link up, interact, and exchange data through digital networks.[10]

Coming Home

Approaching your driveway, a signal from your phone or tablet causes the garage doors to open. Another touch and a door unlocks; and with your voice command, "Hey, I'm home," the wallpaper glows, the home is heated or cooled to your wishes, and a cool or warm breeze follows you around. The television is set to your favorite channel, your beverage of choice is on the coffee table and, as you move around, lights come on ahead of you and fade behind you. Motion and sound sensors will open doors and tell you if anyone else is in the house.

Favorite songs can be programmed to follow you throughout the house; the walls can change colors and photos or play video at will. The system keeps track of all that you do, and everyone can be "pinned" via an electronic tracking chip that makes adjustments as it learns everyone's preferences. When two different "chips" enter the same room, the system tries to compromise, as with music or TV, on something that both people will enjoy.

Acting as an extra energy storage unit, vehicles can be used as buffers to deliver electricity to compensate for energy output or input of a home's energy system. This technology can also be utilized for electronic appliances, to ease electrical flow at peak times and to allow the grid to "borrow" electricity from appliances during times of need.[11]

There are many "peak demand" power plants in the United States that only produce electricity for a minimal amount of time in order to prevent grid overloads during peak load times. These plants are expensive to operate, so there's a great demand for cheaper solutions.[12] Smart homes and other buildings will be equipped with a "smart energy box" that can lower energy demand in a targeted manner during peak load phases, easing the strain on the grid, conserving energy, and saving money by as much as 30 percent, while maintaining the proper level of comfort for building occupants. The future supercity will make energy miners and energy traders of us all. Sensors throughout the city will provide essential information to keep energy use running efficiently.[13]

Some Smart Tasks

The IoT is a system of interrelated computing devices, mechanical and digital machines, objects, and animals or people that are provided with unique identifiers (UIDs) and the ability to transfer data over a network without requiring human-to-human or human-to-computer interaction.[14] Devices now can collect and transfer data about routine behaviors to control centers in the cloud for analysis and then program and/or alter the information for preferred settings and courses of actions.

Some of the most common centrally controlled technologies in today's smart home include:

- Smart locks that provide a new home security experience with the ability to customize who can access your home and when, lock or unlock your door from anywhere with a smartphone, and even unlock the door by voice.[15]
- A smart hub to ensure that smart devices can communicate properly with each other by linking several different products, including speakers, cameras, computers, smartphones, televisions, security systems, appliances, and more.
- Automated heating and air conditioning systems offering the potential to save energy and money with smart thermostats that quickly and precisely automate the heating and cooling of a home, potentially reducing your electric bill. Other products such as connected lights and appliances can use less energy by powering down when not in use.
- Connected lights, cameras, and even doorbells that can help make a home that much safer. If you're home alone and burglars, for example, are checking out your house to see if they can break in, having these kinds of devices can alert authorities or simply scare intruders off.
- Automation of some of the tedious tasks of home care, such as automatic vacuuming, a machine that automatically starts a wash or dry cycle, or a fridge that orders groceries online when it senses that you're low.
- Entertainment and information utilities, which consist of connected speakers with AI-based digital assistants can play music, offer news and sports scores, and recommend a good film to watch either at home or at a movie theater.
- Water sensors to immediately alert you whenever excess water is detected where it doesn't belong. (By the way, according to HomeAdvisor.com, the average cost to repair home water damage throughout the country is $2,476.[16])
- Kid monitoring by smartphone, to easily and effortlessly check in on family members and pets. Know when kids come home from school, when cars arrive in the driveway, or when pets unexpectedly leave the house. For young children, parents, or pets that don't carry a smartphone, simply place a SmartSense Presence sensor in their backpack, purse or pocket, or around their collar, respectively, to know when they come and go.
- SmartThings, which allow you to protect your family by monitoring and securing dangerous and off-limit areas. Get immediate

alerts if children open things like cleaning supply cabinets, medi-cine drawers, or gun cases.

- Med-monitoring bottles that sense or count the number of pills taken and advise on when to order more.
- Ovens that set an optimal time and cooking temperature so that you can check on Thanksgiving dinner while watching the football game.
- Path lighting, for nighttime bathroom trips by voice or by touch-ing a wall, or for instantly creating mood lighting for any occasion.
- Bedroom warming for when you get out of bed, so that it's nice and toasty when you get up.
- Light regulation smart bulbs, based on the availability of daylight.
- TV monitoring, so that your children can watch only approved programs and at certain times.
- Smart security cameras to monitor the home when residents are doing errands, at work, or on vacation. Smart motion sensors are also able to identify the difference between residents, visitors, pets, and unauthorized persons and can notify residents and au-thorities when suspicious behavior is detected.
- Monitors for how much you throw away and what you throw away.
- Pet care utilities for feeding your pets on a schedule with a preset amount of food, cleaning litter boxes and dumping the contents on schedule.
- Plant and lawn watering only when needed and with the exact amount of water necessary.
- Management of battery storage of solar energy and a charge for an electric vehicle.
- Connection to an electrical grid for information about available power and where energy is needed.[17]

THE FUTURE SMART CITY

The future smart city will employ a number of powerful automation and AI programs to process vast amounts of incoming data. These programs will track and upgrade improvements in automation and AI networks immediately. In fact, smart cities may witness the birth of the first truly large-scale AI, capable of interactive, reactive, and independent think-ing to monitor and safeguard a city.

Geostationary and other satellites and orbital platforms will monitor the city's atmosphere, pollution levels, weather systems, and local envi-ronment across the electromagnetic (EM) spectrum, with particular attention paid to potential threats from earthquakes, tsunamis, hurricanes, tornadoes, and other natural disasters.

Urban "stack farms" will put vertical building space to efficient use in producing food for the city's population, conferring on the smart city an unheard-of degree of agricultural autonomy. Integrated transportation systems, meanwhile, will reduce traffic congestion and strongly limit pollution.[18]

Sensors, networks, and wireless systems will communicate information about the health and status of the city and its infrastructure. Satellites will monitor the city's atmosphere, pollution levels, weather systems, resource availability, and energy use.[19]

Sufficient energy to power smart cities will be generated from clean, renewable sources, with each power system compartmentalized for quick isolation with robust backup systems in case of failure—all watched over by a smart system.[20]

These are just a few of the more notable features of the future smart house and city. With more than half of the human species huddling together in dense urban areas, it's inevitable that our cities will need to be monitored and upgraded constantly. Cities of the future will be less defined by their skylines and more by the sophistication and upkeep of their "smarts."[21]

VISIT TO A VIRTUAL DOCTOR'S OFFICE

In various surveys across the country, health care is one of the most important issues. In the future we'll be able to conduct a pre–doctor visit by employing home health-care equipment, such as monitoring and diagnostic tools. Digital developments in health care will also mean fewer visits to the doctor. The process of taking care of millions of adults worldwide will be simplified and will include monitoring, through radio frequency identification (RFID) chips, the movements of someone suffering from dementia, and making sure people get and take their medicine.[22]

Homes will be equipped with body monitors, which will scan and provide basic digital diagnosis of residents. Watches will be equipped with electrocardiograms to detect possible irregular heart rhythms, with which six million people are diagnosed every year. Close to seven hundred thousand are affected by heart problems and don't even know it, the occurrences of which are expected to double by 2030. A smart health system app can detect an A-fib immediately and notify the endangered and the closest health-care outlet.[23]

Smart health systems/software will also recommend various basic treatments and even order medication (with consultation from a doctor or a doctor's personal digital assistant) that can be delivered by an air drone or a building robot. The inclusion of health-monitoring equipment in the home could have a tremendous beneficial impact on average families, especially those in rural areas. For example, a home monitor could check the heart rates of its occupants, track any illnesses or preexisting conditions, check meds, automatically alert doctors or first responders about any abnormalities or problems, and also advise a family member or significant other when a loved one is suffering a health emergency.[24]

Another innovation is a smartphone application called Mind Monitoring Systems. The app is designed to track a user's emotions and state of mind throughout the day using voice measurements.

Google's latest patent describes using ultrasonic bathtubs, pressure sensing toilet seats, and other devices to monitor people's cardiovascular health. The pressure-sensing toilet seat would also be used to measure the patient's blood pressure, as well as monitor the individual's health through their bowel movements.

An ultrasonic bathtub can generate high-frequency sound waves and gather an echo from these waves in order to investigate the user's internal body structure—blood flow and tissue movement—and create 3D measurements of that structure.

Another device is a color-sensing mirror that records a person's skin tone to determine irregularities; it also measures variations in a size or color of an organ or limb—or growth. A radar field device would reflect radiation from human tissue to measure skin temperature, heart rate, and skeletal movement.[25]

Technology will be available that determines trends for human physiological systems in order to catch life-threatening states before it is too late. That includes any abnormalities in your pee and poop that could be signs of cancer and incipient diabetes, among a host of other diseases. The technology will also deliver this information to your personal physician.[26]

UNCOMMON CONDUCTIVITY

A present problem is that very few companies produce all the devices that may be found in a household or know how to connect them all to a smart hub to ensure that the devices will work together efficiently. It's also unlikely that consumers would be loyal enough to buy every household device or even a majority from a single manufacturer. So if companies want to ensure that their devices talk to others, they will have to develop all devices with common standards and shared software. This level of collaboration may take some time, as many of these firms are direct competitors.[27]

Perhaps the next question is, is all this really necessary? Was the auto a necessity at the turn of the twentieth century, or the radio or phone around the same time, the airplane, refrigerators, the national electric grid, television, rocketry and satellites, the personal computer, and the internet? These are a few of the things that people lived without until they were invented and then became things we couldn't live without.

At present, fully integrated custom systems are expensive and often require a consultant to install them and possible structural changes to the home—all involving costs tacked onto the price of the system itself. Systems and their abilities range from moderately expensive to well outside the range of the average consumer. However, costs will lower with increased demand.[28]

DUMB DEFECTS

A big concern is the potential for criminals to hack into a smart-home system, as home security is high on the consumer want list.[29] This has serious implications, as smart-home systems generally integrate with home security systems, leaving not only your home but confidential secrets about you, your family, finances, and other secure information vulnerable. A study by Hewlett-Packard exposed some problems:

- 100 percent of the studied devices used in home security contain significant vulnerabilities, including password security, encryption, and authentication issues.
- 80 percent of devices tested failed password security, with most devices allowing such mundane passwords as 1234.
- 70 percent of IoT devices examined failed to encrypt communications to the internet and local networks, and half allowed unencrypted communications.
- The user interfaces of six out of ten devices tested had issues such as XSS inserts (a flawed code that allows hackers entry), poor session management, weak default credentials, and credentials transmitted in clear text.

- 60 percent of devices didn't deploy encryption with software downloads.[30]

A systems integration report found that 92 percent of consumers were concerned about the security of their homes and smart-home data. If hackers were able to infiltrate a smart device, they could potentially turn off the lights and alarms and unlock doors, leaving a home defenseless to a break-in. Furthermore, hackers could potentially access the homeowner's network, leading to the discovery of all kinds of sensitive material and making one vulnerable to crimes such as blackmail, extortion, or identity theft.[31]

BODYGUARD OR BIG BROTHER

Advanced smart security systems can make a record of and notify you if there has been an intrusion, detect unfamiliar vehicles approaching your home, sense unfamiliar smells and step patterns, notify the authorities, automatically lock doors, provide room-by-room surveillance, and track unfamiliar footsteps, voices, and strangers at the door.[32] Security is an important aspect to most people, and many will take comfort in the idea that someone or something is watching out for the safety of their goods and people. After all, it took the FBI only three days to run down the two Boston bombing suspects, aided by a department store's cameras,[33] and inside of a week for London investigators to identify terrorist's attackers with the aid of street cameras.[34]

But it's a touchy subject to others who will irately scream about governmental abuse and override. The idea of constant monitoring feels un-American to many of us, especially those who laughed when comic Yakov Smirnoff said , "In America you watch television; in Soviet Union television watches you."[35]

A MOUNTING MARKET

While less than 1 percent of homes currently employ full smart-home technology, Allied Market Research believes that the market will grow at an annual rate of 30 percent through 2020, at which point the market will be worth $35.3 billion.[36] "The global smart home market is forecast to reach a value of more than $53 billion by 2022," reports Zion Market Research.[37] The number of devices connected to the internet is estimated to grow by 140 percent, reaching fifty billion worldwide by 2030, according to the latest study by Strategy Analytics. [38]

IoT devices have already outnumbered the human population. With constantly dropping prices of sensors, connectivity, and so forth, the era of smart homes is not very far off. But how useful or widely accepted that era will be is still a topic of skepticism. Whether we all want it or not; we definitely can't escape it. [39]

Driving the market will be the decreasing costs of smart technologies, increased government regulation regarding energy consumption, increased energy costs, improved consumer awareness issues about the environment, and increasing consumer security concerns. [40]

What's This Going to Cost?

The average cost of a smart home is around $1,000 per room, with most homes having an average of five smart rooms. A system bus or hub is a single computer that links all of the devices within a home or network. The basic costs of installing a starter home system range from $40 to $500. Wireless systems run from $300 to $600. And monthly service is $35 to $70 per month, plus activation fees, which normally cost around $200 to $500. [41]

But like anything else, you can spend as much as you are able. So be careful to check out the options for connecting appliances that you already own against brand new stuff off the shelf.

A Bot in Every Abode

Several companies are or will be debuting robot makes and models that will be far more functional than a Roomba but not quite as formidable as *Star Wars'* C-3PO. Many experts and futurists predict that in the next several decades, robots will be in every household. They will likely be fully integrated with the smart-home operating system and will help manage it. They will not only assist with manual tasks but will also learn a family's moods, patterns, preferences, and behavior and become a teacher, helping with homework or other problems. [42]

Robots will be programmed to recognize faces and respond appropriately to the emotions of people. Educational robots may also target more advanced learners and will soon be online teaching. Their facial recognition capabilities would be used to shake things up when students look bored or are unmotivated, and they may even take a role in parenting by reading bedtime stories and helping with social problems. They will soothe crying kids, play or sing to them, be equipped with baby monitors and built-in cameras, and even act as nightlights. Bigger, beefier bots will be designed to handle heavyweight duties and carry out tasks like lifting and carrying. [43]

Diving into the sub-rosa world of sex, a number of companies are currently developing robots designed to combine utility and physical pleasure–sexbots with a few already on the market. Unlike sex toys and dolls, which are typically sold in off-the-radar shops and hidden in closets, sexbots may become mainstream. A 2017 survey suggested almost half of Americans think that having sex with robots will become a common practice within fifty years. [44]

Like anything else, this synthetic coupling comes with a price tag that can range from thousands to a place in Dublin where for an hour you can indulge in silicon sex for about 88 pounds, or about $100. [45] And apparently, even though sexbots are for sale in the United States, it will take time for morality and the laws to catch up with human desire. But be nice, because when the machines rise, as they do in the science fiction films, you may not want to have anything out of the ordinary to answer for.

PROS AND CONS

As a recap, consider the following smart-system balance sheet.

Pros

- Peace of mind. Smart homes are well-known for their improved security. Many systems come with remote dashboard capabilities, so forgetting to turn off that coffee pot before you leave no longer requires a trip back to the house. Simply pull up the smart dashboard on a smart device or computer and turn the coffee pot off in a matter of seconds.
- Efficiency. When a system obtains information from cars, appliances, or energy grids regarding their requirements, it creates an energy demand forecast for the next day.
- Accountability of designers and builders. They will become more accountable for increasing the efficiency of energy use and for managing the resource demands of buildings during construction and remodeling, along with providing consumption forecasts for the life of the buildings and reducing negative impacts on human health and the environment.

- Savings. A study by the American Council for an Energy Efficient Economy found commercial buildings could collectively save up to $60 billion by increasing energy efficiency investments by just 1 to 4 percent. Medium-sized buildings can expect to save up to 25 percent on energy costs with smart energy management, and returns can happen quickly with very little oversight and maintenance from staff.[46]
- Ability to customize as many or as few electronic devices as you choose. They learn and adapt to your preferences without your having to ever input a preselected schedule. Either traditional or behavior-based automation can be applied to virtually every gadget that can be remotely controlled, from sprinkler systems to coffeemakers.
- Wirelessness. Smart-home technology and compatible devices are becoming more friendly, and many can be installed with a minimum of tools and expertise, using only the information provided in the owner's manual.

Cons

- Home smart systems have struggled to become mainstream and have suffered in part due to their technical nature. There may be a learning curve for non-tech-savvy people.
- Depending on the complexity of the system, installing a home automation device can be a significant burden on the homeowner. It can either cost you money if you hire an outside contractor or cost you time if you venture to do it yourself.
- Not all systems are compatible with one another. Your security system, for example, may require you to log into one location to manage settings, while your smart thermostat may require you to log into another platform to turn the air conditioner on and off. To truly leverage the convenience of home automation, you may need to invest in a centralized platform to control all systems and devices from one location.
- Smart-home technology can require not only an investment in a centralized platform but also expensive add-ons that will replace existing fixtures that expire or may become obsolete. For example, light switches and controls may need to be changed out from basic light switches to "smart" controls that are able to accept input from smart programs. A standard light switch may cost a buck or two while a "smart" light switch might cost up to $40.[47]

SMALL AND SMART POWER GRIDS

Efficient, flexible, and interactive electrical smart grid technologies will shift energy management away from a centralized producer-controlled utility network to a smaller decentralized, proactive, and self-controlled grid from city to city or building to building. This will turn your power grid into a demand-controlled tool, enabling you to generate, deliver, save, and store power more efficiently.[48]

Automating everything in life may sound extremely appealing, but sometimes a good old-fashioned flip of the switch is a lot easier than reaching for your smartphone to turn lights on and off. Before you decide which system is right for you, think about how far you want to take home automation in your home and how you will communicate with it.

12

SUPER SKYSCRAPERS

How High Can We Go?

Skyscrapers are the rock stars of buildings.

—Unknown

What was fiction in building big buildings yesterday is science today. And the science fiction of today will be the reality of super skyscrapers tomorrow. In a skyline they make a municipality look "big city." And activity with huge cranes and buildings climbing upward can often mark which cities are booming. In the future they will soar into the sky, float in and under the seas, burrow underground, even orbit the earth, and will be humankind's mark on alien landscapes.

These edifices will be powered by sun, wind, and tides; keep the air inside clean; use and store clean recycled water; employ a myriad of innovative means that produce electricity to run smart cities; and recycle to net zero.

Foods will grow around, in, and on top of buildings and in indoor hydroponic gardens. Three-dimensional printers and modern materials will be used to create modular homes, parts, and buildings made robotically in days instead of years, at a cost of thousands compared to millions, and millions compared to billions.

A future vision for the urban skyscraper is one that will combine offices, public spaces, energy sustainability, mobility, climate adaptation, water management, green space, food production, and resident comfort in a self-sufficient structure. An artificially intelligent smart system will run the operation, management, and maintenance and will function as an inclusive organism. And the structure will be one that in the future, from birth to death, you may never have to leave.[1]

Skyscrapers will have to be more than just tall buildings; they will need to be denser, taller, relatively self-contained, and autonomous. They will have to have the ability to adapt to their surroundings and with their smart systems become multifunctional, high performance, and self-sustaining. The next skyscrapers may not look like current skyscrapers at all—they'll still be vertical cities, but they'll take on whatever form is necessary to best provide for their occupants, with a toolbox of systems that will respond to the needs of their inhabitants whether individually or collectively.[2]

These high-capacity, high-efficiency, ultratall buildings will probably occupy a relatively small, car-free, pedestrian-friendly parcel of land. Within this footprint are all the self-sustaining features, facilities, and services necessary for satisfying and improving the living, working, cultural, entertainment, recreational, and leisure essentials of living for residents.[3]

A SHORT HISTORY OF TALL BUILDINGS

For thousands of years the tallest structures erected were typically only church spires. But as the technology of the Industrial Revolution expanded, so did the places people worked; and as the population expanded, so did the use of multistory buildings in cities at the end of the nineteenth century. Technology at the time was advancing quickly, but people were still freaked about the safety of such tall buildings, and height limits at the time were fairly common in American cities. So, for example, the Height of Buildings Act of 1910 was passed by Congress on June 1, 1910, to limit the height of buildings in the District of Columbia to a range of 90 to 130 feet.[4]

Even though the law was never repealed, changes in urban life encouraged the switch to taller, higher-density facilities.[5] Street trams, subways, and elevated rail links provided the means to deliver hundreds of workers to a single urban location, and soon afterward structures started to stretch skyward as the ranks of workers flooded the cities.

In the 1930s, within the span of just two years, the world's tallest building was built three times in New York City: first, the 827-foot Bank of Manhattan in 1930; next, the 1,047-foot Chrysler Building in the same year; and last, the 1,250-foot Empire State Building in 1931, which was the tallest in the world for two decades.[6] Nowadays it seems that skyscrapers get topped every couple of years. But the idea of ultramegastructures is not new. Frank Lloyd Wright's 1956 proposal, The Illinois, was a projected mile-high skyscraper. The design included 528 stories, with a gross area of 18.46 million square feet.[7]

The era of architectural design and erection has been a kind of an arm wrestle that has only intensified since its beginnings. In 2003, the 1,670-foot Taipei 101 unseated the 1,483-foot Petronas Towers in Kuala Lumpur; and in 2010, the Burj Khalifa in Dubai increased the climb to 2,716.54 feet. Bold builders in China have proposed a 220-story prefab tower and claim it can be constructed in an astonishing ninety days.[8] Recently proposed skyscrapers have been projected for heights of a mile and higher.[9]

For much of the twentieth century, the one hundred tallest skyscrapers stood in North America, until Asia began building tall towers in the '80s and '90s. Today the Middle East and Asia are home to most of the world's tallest skyscrapers.[10]

SUPERSTRUCTURES

There's something mysterious and aloof about skyscrapers—they seem to defy gravity, and their builders have to include the curvature of Earth, defeat high winds, calculate incredible loads, and ensure suitably strong, stiff, and robust stability systems for both the interior and exterior. And whether you are looking up or down from skyscrapers, they are dazzling and dizzying.

They also cause a little city chest pounding—who can forget the symbolism of King Kong atop the Empire State Building? Every city wants to be the place for the tallest, the most unique, and most innovative skyscraper; a kind of competition transpires between builders, architects, and metropoles. A soaring skyline is also a high watermark of economic boom in a metro area.

How Tall Is Highest?

Theoretically the tallest building could throw a shadow on lofty Mount Everest (29,029 feet).[11] Buildings are only 15 percent as heavy as a solid object—far lighter than a mountain. According to the Skyscraper Society, a building can be 6.6667 times taller than a solid object.[12] Quick math indicates that theoretically a building could be far higher than Mount Everest, at a little more than 193,000 feet. Of course, such height would raise other, quirky issues. Its base would have to be about 2,858.307 square miles—a rather large lot for any city.[13] Then there is the problem of planes making uninvited crash landings and issues related to oxygen and sunlight at that height. It sounds like a long elevator ride too.

Getting Up

Without elevators multistory buildings wouldn't exist. By the way, Elisha Otis didn't invent the elevator; Archimedes is believed to have built the first one 2,200 years ago.[14] And in the eighteenth century, Louis XV is said to have had a personal lift installed at the palace of Versailles so that he could visit his mistress.[15] But Otis did invent the safety brakes in 1854, allowing the trip up and down to occur in relative security, which was the gateway needed for the modern high-risers.[16]

Soon maglev motors and lightweight materials will allow elevators to go side to side as well as up and down. Designers speculate that buildings implementing this technology could increase the carrying capacity by as much as 50 percent.[17]

Each car will be powered by two magnetic linear maglev motors, one for vertical and another for horizontal movement, and equipped with light but enormously strong carbon-fiber cables, cutting out steel cables and winches in order to reduce weight. New-generation maglev elevators will enable the shattering of height limitations as well as introduce a new mode of side-to-side, building-to-building transportation, since horizontal elevator shafts can be built into neighboring buildings.[18]

They also might be powered by compressed air so that a ride from the ground level to the topmost floors would only take only a few moments—at a speed that might make some passengers dizzy or their stomachs woozy. Elevators might also run snakelike, like a vertical, circularly stacked train long enough to service as many as thirty floors at once—and you'll only have to wait seconds for a seat.[19]

The fastest elevator at present is in the CTF Financial Center skyscraper in Guangzhou, China, that will travel at about sixty-six feet per second—a pace that might be troubling to popping eardrums and the decelerating feeling of weightlessness in the guts of passengers.[20] It'll be almost as fun as a roller coaster.

Getting Skinny

Tall, skinny apartment towers are on the rise, sprouting like magic new-age beanstalks from small lots—some only as wide as a handful of townhomes. Fueling the drive toward slim living are factors led by a robust demand from the high end of the residential brackets, people willing to pay big bucks for status and outrageous views. Forward thinking in structural design and bleeding-edge building materials have made constructing skinny skyscrapers possible. It makes sense to developers that if you can build slender and higher, you can get more and higher units with an elevated price tag.[21]

* * *

The skyscraper of the future isn't just about scraping or even piercing the sky but also about making its space "high performance."[22] Mega-metro markets will not only have to deal with living and working spaces but also resources—water, energy, air, waste, and food, transportation, quality of environment, citizen services, building methods and materials, and ways to deliver these products and services that don't take a big-time bank account. In other words, neo-skyscrapers have to be and do more than just go upward. They'll have to integrate smart technology and become autonomous, multifunctional organisms in their own right.

What it will take is not only careful planning but also looking to a variety of alternatives and solutions, some of which will sound reasonably practical and doable and others like questionable new-age, sci-fi futurism.

To accommodate the oncoming stampede of a hyperdense Gotham City, for example, superbuildings of the future will consist of millions of inhabitants, and they will provide a framework for just about everything in a single mammoth structure that could soar thousands of feet into the sky and spread several city blocks wide. It would offer quality living, affordable housing, and an accessible, clean out-of-doors experience, where the air is refreshing, the food fresh and sustainable, the water pure, and energy ample and inexpensive. Single-stream recycling will recover and recycle just about everything, and buildings will generate net plus energy (more than they consume) and stash the surplus into storage or pass it on to other mini or regional electrical grids.[23] Even the temperature of the building, plus the heat of generating electricity, will be captured to generate more electricity.[24]

Food would be supplied by local producers, public transportation would be simple and affordable, and all necessities would be nearby and easily available. A supercity skyscraper would not only be a pleasure in which to live but would also provide living spaces that would be attractive and easier to organize, administer, operate, and live in.

ALL SHOPS, ONE STOP

This is what an "everyday" day in a supercity skyscraper might look like. Get the kids up and down to the lobby, where their transport to school awaits unless instruction and schools are in the building. Stop for a coffee in the lobby or have one sent up via your electric dumbwaiter. Perhaps go to a doctor's appointment at the clinic on an upper floor or a little shopping on the floor where the mall and supermarket are located. Afterward take a stroll in the building on your favorite nature trail, check with your phone to see if your purchases have been delivered via a drone or your homebot, and then meet the kids in the lobby diner for an afterschool treat. After preprogrammed entertainment, dinner, and AI assistance with homework, the children are in bed, and you and your partner step out to the bistro on the top floor for an after-dinner cock-tail and live entertainment, which you saw previews about on the build-ing intranet channel. Check in with your babysitting robot, and consult your schedule, which shows options for everything from food to travel, your exercise plan, weekly calendar for the family, and interesting and upcoming things each family member might be interested in—all with-out putting a foot outside your building. Even disturbances such as natural disasters will be planned for in every aspect—even though you'll be thoroughly protected in your building.

   In the "tomorrow land" of major cities, residents in superstructures will have their own networks for shopping, recreation, medical facilities, entertainment, schools, inner- and extra-city transportation, foods fresh off the urban "farmlands," and parks—all concentrated under one roof within a single massive structure, stretching into the sky and consuming many city blocks in width. These superbuildings are close to becoming a reality because there are planners who believe these megastructures will be our best option for dealing with the future demographic, eco-nomic, and environmental onslaughts that are right around the corner for cities.[25]

IMAGINING EDIFICE REX

Visionaries are putting efforts into raising superstructures higher and higher, so that soon we will live and work among the clouds or in the darkness of space. Today's skyscrapers scarcely scratch the sky com-pared to those of the next couple of generations. Tomorrow's will be built perhaps in orbit or on asteroids and anchored to complete ecosys-tems unto themselves.

One chimerical structure, architect Eugene Tssui's still unbuilt Ultima Tower, would comprise 500 stories; likely have a floor area of about 1,500,000 square feet, the interior of which would contain 39,000 acres; and cost an estimated $150 billion. It would have a stable, aerodynamic trumpetlike or bell-curve structure and accommodate up to 1 million San Franciscans.[26]

For a foundation, such a structure might have reinforced geopolymers (a concrete that is made by reacting aluminate and silicate-bearing materials) with a metal microlattice core that is one hundred times lighter than Styrofoam but far stronger and far less expensive than titanium. It could be wrapped with a double-helix carbon-fiber skin that has the strength one hundred times that of aluminum but with far less weight. The exterior walls could be made of structural glass, virtually bulletproof, designed to disperse all forces along the surface.

By the way, as far as proposed structures go, the tallest would really have to be the hypothetical space elevator. It's planned as a 328,084-foot carbon-fiber structure with the shaft anchored on Earth that rises beyond our atmosphere, where a counterweight would hold it in place, enabling Earth-based vehicles to be delivered into space sans booster rockets. It would also be used as a platform to deliver supplies to space colonies and travelers to astro-hotels, and as a tourist stop (see chapter 5).[27]

INDIVIDUAL ECOSPHERES

Don't think floors, but imagine entire landscaped neighborhood "districts" that are fifty to one hundred feet high, each with its own mini-biosystem featuring small lakes, hills, streams, natural sunlight, fresh air, all manner of greenery, and panoramic views. It's of primary importance to bring nature upward to promote and enhance natural surroundings in a controlled environment, as if nature grew upward into the edifice. Gardens will be situated at all exterior and interior openings. Terraces, sunshades, natural ventilation, and integrated green space will be designed to bring in light and space to living areas by means of reflecting mirrors when necessary, and with either personal or automated controls of ventilation of outdoor or indoor fresh air (see chapter 6).[28]

Green rooftop gardens, with birds, small animals, and fish that are brought together to mutually benefit one another as well as economically produce fresh foods through aquaponics will be the setting (see chapter 10).[29]

Water will be placed via bladders on specific levels; the bladders will serve as fire barriers, sprinkler system reservoirs, supplementary hot- and cold-water holding tanks, and catch basins for rain for lakes, water- falls, and streams to help support whole ecosystems within the building. Water systems for residential needs will be supplied by various meas- ures from natural sources and refiltration devices, such as toilet-to-tap systems and desalinization where possible—nothing will go to waste.

Interiors will feature a new type of wallpaper that can change colors and patterns manually or under certain light.[30] Tired of looking at a San Francisco vista or the *Mona Lisa*? Try Rome or Picasso instead. Bored with the carpet color? That, too, will change at the sweep of a hand or push of a button. Whole walls will be used as TV screens in any room, if desired.[31] Residents will have access to additional communication and RFID devices (radio frequency devices that can identify and find any- one, anywhere), and a range of communication mechanisms that can spread news through the city via billboards, street signs, and Siri/Goo- gle/computer signals.[32] Buildings will be made from self-cleaning mate- rials, and for tough spots sanitary robots will be available.

Building towers themselves may be surrounded by water featuring sandy beaches, stone cliffs, plants and trees, small islands, birds, and squirrels and other small animals. Lake water will be drawn up through- out the structure and used for cooling floors and walls as well as for emergencies. A portion of this water will be heated by large, passive solar panels and left to descend by gravity, to be used at various levels.[33]

Such a building/ecosphere will feature several aerodrome ports for drones to deliver packages and passengers.[34] Below the surface level, there will be a terminus for underground maglev or superspeed electri- cal subways connecting to inner- or extra-city terminals and long-range transportation facilities.[35] Only electric vehicles, people-powered vehi- cles, and some propane and hydrogen gas vehicles for heavy or large loads will be allowed. No gasoline internal-combustion engines or toxic pollutants via petrofuels will exist within the confines of the city.[36]

In "muni-parks," pedestrian walkways and running and cycling paths will meander through hills and the natural landscape. Small green pe- destrian bridgeways will connect buildings. All residential building "neighborhoods" will be located at the outer edge of the building, clos- est to views and natural sunlight. The core of the building will be re- served for commercial use and support systems like schools, churches, clinics, and stores.

HVAC: Heating and Cooling

Cooling (and heating) will be based on water, in this case waterfalls: the cool air will be warmed in the upper floors and exit through different levels of the building, cooling as it falls back to the ground level, either to be cooled underground or by stored energy.[37] To enhance this effect cooling bricks will be used that absorb water and chill air is it passes over and through the bricks. Winds will be also cooled as they flow through the spaces and bladders of water between the building's cooling or warming ventilation.

All windows in the building will be both electrically and manually operable and act as natural air conditioners. Building HVAC systems will follow the sunlight over the course of the day, sucking dew and carbon dioxide out of the air. These substances will be filtered and stored.[38]

Deterring Natural Disasters

When planning for ultralofty buildings, natural disasters must be taken into consideration. For example, in San Francisco one has to think of earthquakes; in the Southeast hurricanes and tornadoes. Buildings must be extremely aerodynamically efficient and resistant to earthquake shock waves, so that if earthquake shock waves push or disturb one portion of the structure, another portion absorbs and dissipates the forces.

A core of metal lattice and carbon fiber in a double helix configuration will keep an entire building in constant tension, producing an equilibrium of stress and strain coming from any direction. Even in a hurricane-force wind or an earthquake, the building will not buckle or become dislodged because of its inherent strength and ability to mitigate or dissipate pushing and pulling forces.

Other ingenious solutions include windproof skyscrapers, like the Taipei 101 building in Taiwan, which is topped off with a giant pendulum that swings in the opposite direction to the wind when a typhoon strikes.[39]

13

NEW USE FOR REFUSE

Gold in Garbage

Human society sustains itself by transforming nature into garbage.
—Mason Cooley

Even the word *waste* leaves a bad taste in our mouths. And it's becoming a misnomer. Waste in the future won't be looked at as something that has to be buried, burned, or carried away. On the contrary, it will be a significant commodity for a valuable makeover in an innovative multibillion-dollar industry that will cause a new-age "waste rush."

Take Los Angeles, for example. The amount of trash created in Los Angeles County is staggering: 10 million residents generate about 14 pounds of garbage per person per day, or 85,692 tons.[1] There's a lot of use for that refuse transforming it into a raw resource.

By 2025, instead of "disposing" garbage, waste companies will morph into a "reprocessing industry," where their central role will be not to dump, burn, or ship stuff off but to extract, repurpose, and recycle everything valuable, returning those resources for reuse to manufacturers, recyclers, or the general public.

A similar redo is required of designers and manufacturers of goods whose after-market refuse should be seen as the raw material of tomorrow. Some are calling for incentives—a positive for some manufacturers who get it but a negative for those who won't get with the program. When our mindset changes, manufacturers will begin to make products that last longer, use more recycled materials, and make products easier to repair, unwrap, and, in the end, dismantle.[2]

DEALING WITH WASTE

A future city that makes no distinction between waste and supply will be caught in the backwash of an endless and problematic supply stream of rubbish. The new green approach will be that there is no such thing as waste, just about everything is reusable, and we should stop looking at waste as something that should just "go away." Making something go away used to mean getting rid of it somewhere else—shipping it out to other locations, burying or burning it, or using any other (unsustainable) way to remove it from sight and smell.

Tomorrow's waste management companies will undergo a metamorphosis into an industry that examines, invents, or researches constructive uses for our throwaways. Their central role will be not to dispose but to return the used to be reused.

In the next few years public attitudes toward waste will require a radical makeover. Half of the food produced around the United States ends up in the garbage can;[3] those leftovers and all manner of material from clothes and paper to plastics and metals to medicine will be thoughtfully moved to incipient supply stream. (Check out my book *Throwaway Nation*.)

Truly sustainable cities of the future will not differentiate between waste and resource. Rather, they will understand waste as the starting point for something new. Ideas and initiatives are taking shape that provide a glimpse of how we could build our urban environments more sustainably in the future.

WRECK WASTE

Roughly 40 percent of solid waste in the United States derives from construction and demolition. Every year, more than 548 millions of tons of construction and demolition waste like timber, concrete, metals, glass, and asphalt end up in landfills in the United States—about double the amount of waste picked up by garbage trucks every year from homes, businesses, and institutions.[4] Such waste involves a significant loss of valuable materials, metals, organic materials, and energy. Thus, there is a great opportunity to create closed material loops in a circular economy. In the future all of the material used in buildings and other structures will be recycled to their origins or into new types of construction material.

According to the *International Business Times*, "Industrial waste from demolished buildings is damaging our environment, but with innovative 3D printing, we are able to recycle construction waste and turn it into new building materials, greatly reducing construction costs."[5]

NEW LIFE FOR URBAN ORE

Landfills will gradually become the mines of the future. Natural resources required for the production of construction materials, like sand and gravel, are depleting, but they stand in huge piles in our urban wastelands.

The Cleveland-based firm of Redhouse Studio has developed a biological process to turn wood scraps and other kinds of construction waste like sheathing, flooring, and organic insulation, into a new, brick-like building material.[6]

The company wants to use the waste materials from the thousands of homes in Cleveland that have been demolished over the last decade or so as a source to create this new biomaterial, which will be contained and recycled in shipping containers repurposed into mobile labs called the Biocyclers.

MONEY IN MUNICIPAL WASTE

People in urban areas produce about twice as much garbage as people in rural areas, and this statistic is only going grow as more and more people migrate into cities.[7] By 2050, metro areas are expected to see an influx of approximately 2.5 billion people and all the garbage they generate.[8]

The good news is that it's possible to take advantage of that refuse in smart ways that benefit a city's energy resources. Landfills produce methane and carbon dioxide, which can be collected and burned for energy or building materials of the future; solid waste can be cleanly incinerated, generating energy.[9] This isn't the cleanest way to rid ourselves of discards, even if the upside is producing energy, but researchers are working hard to create a carbon capture system to collect excess emissions, pack them back into the earth, or use them to create new materials.

Only about one-third of our greenhouse gas is actually turned into electricity.[10] The rest of the gas is either flared (burnt off) or isn't recovered at all and just floats off into the atmosphere. The largest methane emitters are oil and gas, agriculture, and waste management.[11]

Clearly there is a lot of room for growth in using these gases for building or energy.

Landfills should be considered as long-term assets rather than negative eyesores. The Organisation for Economic Co-operation and Development (promoting policies that will improve the economic and social well-being) states that three billion tons of trash a year will contribute to landfills worldwide by 2030, up from 1.6 billion in 2005.[12] However, the potential of landfills as a vast new resource, as opposed to a useless burden, is encouraging.

The concept of "by-product synergy" consists of taking the waste stream from one production process and using it to make new products.[13] Using waste instead of trashing it can cut costs by reducing disposal fees, lessening the use of virgin resources, diminishing greenhouse gases, and opening up additional revenue streams through by-product sales—thereby creating a new industry: waste mining.[14]

Losing Livestock and Plant Weight and Waste

To produce a pound of beef takes 12 pounds of grain, 2,500 gallons of water, and lots of grazing land.[15] Roughly 20 percent of all currently threatened and endangered species in the United States are harmed by livestock grazing. Animal agriculture is a chief contributor to water and air pollution. America's farm animals produce ten times the waste produced by the human population.[16]

Agricultural waste like corn husks or scraps from dining halls can cut the cost of animal feed by reusing (looping) it for feed. Another example of reusing waste is in the cement manufacturing industry. Slag (waste) from a steel mill is used as a material for 3D buildings while also decreasing nitrogen oxide emissions.[17]

ABUNDANCE OF E-WASTE TO MINE

Many millions of tons of televisions, phones, and other electronic equipment are discarded each year, despite them being a rich source of metals. In fact, according to a study in the journal *Environmental Science & Technology*, a gold mine can generate five or six grams per ton of raw material. However, that figure rises to as much as 350 grams per ton when the source is e-waste. And with no huge mining machines to buy and maintain, no toxic waste, and no environmental damage, it's an environmental win.[18]

The amount of e-waste being generated certainly suggests that the business opportunities for recycling could be great. The United Nations International Telecommunications Union estimates that about forty-five million tons of e-waste was generated in 2016, and the amount is expected to top fifty million tons by 2021.[19] Besides, we will have no choice in the future but to take care of our own messes, as our biggest junk dealer, China, is now cracking down on imports of e-waste. There's increasing pressure on the United States and other countries to find more proactive solutions for their own refuse.[20] E-waste mining has already shown the potential to become big business in Australia. Economic modeling shows the cost of around $500,000 for a microfactory pays off in two to three years, generate revenue, and create jobs.[21]

Some of our industries have been criticized for being intractable with the recycling of their products. Apple is trying to counter the opinion that its phones are difficult to recycle with the recent demonstration of a robot, "Daisy," that can disassemble up to two hundred iPhones an hour.[22] But they are still not made to be fixed.

Cell phones alone contain as many as sixty elements, including rare metals such as iridium, which is used in touchscreen technology.[23] To put numbers on it, the EPA states that for every one million cell phones recycled, we can extract amounts of irreplaceable precious metals that can't be ignored:[24]

- 35,274 pounds of copper, worth $423,288
- 772 pounds of silver, worth $9,264
- 75 pounds of gold, worth $216,000,000
- 33 pounds of palladium, worth $396,000[25]

In the United States more than half a billion cell phones are ready for recycling, more than twelve million phones are added to that total each month, and almost 150 million mobile phones are discarded each year. Mobile phones contain numerous metals, including expensive ones. The most important are copper, nickel, silver, gold, platinum group metals, cobalt, lithium, lead, tin, zinc, gallium, indium, iron, chromium, niobium, tantalum, and titanium. Only 25 percent are recycled. It is estimated that the recycled metal market as a whole will be worth $476.2 billion in the United States by 2024.[26]

COLLECTING AND CONTAINING

Cities are also becoming increasingly aware of the problems of waste collection and are contemplating using a small device equipped with a fill-level sensor that will be mounted inside trash containers, to continuously monitor the waste level at pickup times. The readings are sent wirelessly to a proprietary waste management platform to be further analyzed by smart systems to keep track of what's thrown away.[27]

In future supercities there will be strategically placed pneumatic waste collection points for mixed waste, organic waste, paper, and metal waste accessible twenty-four hours a day.[28] After being deposited in "waste inlets" located by your home or in other buildings, garbage will be transported along large-diameter steel pipes that are hermetically welded, and transferred into containers that are sent away for further processing using the city's existing underground railway network. Sealed biowaste tanks will be used instead of plastic containers to ensure that high levels of hygiene are maintained throughout the system, decreasing the potential for food for vermin. This process will be remotely monitored and controlled by smart systems at waste stations. No trucks will be needed, which will reduce traffic and therefore lessen greenhouse gas emissions along with aroma and noise.[29]

The entire network will be monitored and controlled underground by a smart system that will be part of the overall smart system of the city.[30] Waste can be tracked, monitored, and calculated for billing by RFID (radio frequency identification) tags to identify customers' waste handling costs based on their actual usage.[31] This will be fair to clients who use far less waste space than their neighbors do.

There will be no unsightly piles of waste or unpleasant odors beneficial for the cleanliness and image of the area. Another benefit is that a pipeline-based waste collection system is very flexible and should not get congested even at peak times. The use of both suction and pneumatic pressure in a system allows for blowing out blockages and using far less energy overall.[32]

MICROORGANISM MINERS

One day landfill sites may be mined for valuable metals using genetically engineered slugs or repurposed microorganisms. Dr. John Collins, SynbiCITE's commercial research director, calls biomining a thing of the near future and believes revolutionary cell-splicing technology called Crispr-Cas9 could herald the ability to create organisms that digest waste and convert it into useful products, or that produce cells designed to change color on contact with certain metals.[33]

Picks and axes will be passé to miners of the future. Techno tools will be smart that will direct robots with sophisticated detection equipment to unearth materials ranging from precious metals to pockets of gases for energy collection. These tools will also include organisms with manipulated and enhanced genes that can detect and absorb precious metals and direct mining operations for rare materials. For example, scientists have created—by accident—a mutant enzyme that breaks down plastic bottles. The breakthrough could help solve the global plastic pollution crisis by enabling for the first time the full recycling of bottles. The new research was spurred by the discovery in 2016 of the first bacterium that had naturally evolved to eat plastic, at a waste dump in Japan. Scientists have now revealed the detailed structure of the crucial enzyme produced by the bug.[34]

Large quantities of lithium that could be mined and reused to create batteries for electric cars lie buried deep within old landfill sites and could be reclaimed with the help of genetically engineered organisms like bacteria biominers—there are already people doing that on a very small scale in gold mines.

Professor Richard Kitney, also a member of the SynbiCITE team, says developments in synthetic biology could allow new types of "plastic-eating biological devices" to be created that digest nonbiodegradable plastics and return biodegradable material.[35]

PRECIOUS METALS FROM POOP

Researchers at the US Geological Survey along with scientists at Arizona State University have measured gold, silver, platinum, copper, zinc, and other precious industrial metals found in biostools (poop).[36]

But it's not just about "poop mining." Gold can be derived from digested food products or dental fixtures. Gold and silver are used to treat arthritis and cancer and also figure in some surgical and diagnostic procedures. Little flakes of gold and silver from jewelry can enter wastewater when a person does the dishes or takes a shower, and precious particles that are used in a variety of consumer products due to their antibacterial properties go down the drain and are flushed out at wastewater treatment plants.[37]

Concentrations of some metals in biosolid material—say about one part per million of gold, for example—can be found in natural occurrences.[38] There's money in sewage treatment pools and waste rock near mining sites where piles of waste are left behind. This waste rock and drainage water could contain metals with concentrations that were too low to be economically recoverable at one time, or metals that weren't of interest then but now have new, high-tech applications and can be mined with ultramodern innovations.[39]

ROADS TO RICHES

Every day on the road cars eject particles of platinum, palladium, and rhodium from their catalytic converters. More than $98 million worth of precious metals accumulate on British roads every year, making their roads a new-age mining opportunity—and you don't have to go underground, as the stuff is sitting on the surface, just waiting to be collected. The UK has about 240,000 miles of paved roads,[40] and the United States has 4.12 million miles to explore.[41] Do the math.

FRESH FUTURE FOR AN OLD NEED

Landfills might stay open for another twenty or more years, as they are still clearly needed, but they can't survive solely as landfills anymore. The website Waste Dive has been tracking the effects of China's import policies in all fifty states, and changes are expected in many aspects. Operators, like one in Sacramento, California, are considering developing a green business park around their landfill and doing recycling as a secondary function.[42]

Technology and the modern systems and practices it brings will improve waste management considerably:

- Robots are rapidly breaking into all areas of the resource-recovery industry, from curbside collection to general resource management. Advanced robots will be able to detect various types of waste and take action to effectively separate and segregate waste.[43]
- Consumers will go postal and start mailing back all kinds of materials for recycling—even tiny items like cigarette butts and cosmetic containers. This will be important as curbside collection becomes less common.[44]
- Wastewater treatment will be on the rise. Solid waste isn't the only type of waste that can produce hydrogen. Wastewater contains plenty of raw material that can be turned into fuel. And in

2017, more than 80 percent of the world's wastewater entered the environment without any filtration or disinfection, according to an estimate by the United Nations Educational, Scientific, and Cultural Organization. At present, the wastewater treatment that does take place consumes 2 to 4 percent of all electricity produced.[45]

SPACE TRASH

Some ask, Why not just blast the trash into space toward the sun in the future, either to lose it in the infinity of space or to let the sun vaporize it? Well, we have too much space junk in near orbit already, but let's take a look and see what the logistics are.

Let's start with a reasonable, down-to-earth figure of $200 million to launch a rocket with a payload of around 15,500 pounds just into Earth orbit.[46] Derek Thompson reported in the *Atlantic* that the world makes an estimated 2.6 trillion pounds of garbage per year and we would need more than 168 million rocket loads chock full of trash to get rid of it all.[47] That would cost $33,696,200,000,000,000 ($33 quadrillion)—but maybe only $16.5 quadrillion if Elon Musk's SpaceX rocket could cut the bill in half. And to get all that trash out of Earth's orbit and on its way to the sun, you'd have to increase the cost by ten times, or to $160.5 quadrillion.[48] Not even Musk, his billionaire buds, and all the people on Earth have that kind of garbage money.

RATCHET UP RESPONSIBILITY AND RECYCLABILITY

Responsibility for retail-related recycling shouldn't fall entirely on consumer shoulders. Makers and retailers that sell unrecyclable packaging should also make a change and be accountable for those products. They need to step up and take responsibility for financing the collection of nonrecycled postconsumer packaging. Less than 14 percent of plastic packaging is currently recycled in the United States.[49]

The United States Industry Association is taking the lead in providing grants for sorting-facility upgrades, to enable the collection of containers used for liquid products such as milk and juice and for polyethylene-coated paperboard or other board and foil laminates for liquid packaging, like the Capri Sun containers.[50] However, Capri Sun and Kraft, despite their promises, have made no significant move toward making their yearly 1.4 billion containers more recyclable.[51]

Some fast-food, beverage, and consumer packaged-goods manufacturers have to have their feet pulled closer to the fire and become actively involved in becoming part of the solution rather than the problem. And if they fail to act, then sanctions and mandates should be enacted, including consumer identification and product boycotting.

In 2014 there were 1,956 waste landfills, down from 6,326 in 1990, according to the Environmental Protection Agency.[52] The marked decline leads to the question of what will happen to disposal companies in the future, and what, if anything, will take their place.

As far as plastics are concerned, closely identifying one plastic polymer from another is critical, and a new infrared spectroscopy system could present an answer. Based on diffuse reflection, the technique enables unique polymer compositions to be distinguished based on their spectral differences, making recycling easier.[53]

Unfortunately, one of the new wonder materials, carbon fiber, which will be used extensively in the future, is typically not biodegradable or photodegradable. While a car part made of carbon fiber won't deteriorate over its useful lifetime, it also means that should the product crack, break, or just no longer be desirable, it won't decompose in a landfill like other materials. And burning it creates toxic fumes.[54]

CHARGING CONSUMERS FOR FOOD WASTE

Half of the food produced around the world ends up in the garbage bin, and many think that consumers should be charged for the food waste they produce. This approach has been successfully tried in Seoul, South Korea. Residents are given cards that include a chip holding the name and address of the cardholder. They scan their identification card, then dispose of their rubbish in a smart bin with a weighing scale, and are simply billed for the corresponding rubbish.[55]

WASTE TO ENERGY

Some operators have set up waste-to-energy facilities to turn waste into power. The global market for turning rubbish into electricity is expected to reach $37.64 billion by 2020.[56]

While most of the growth to date has been in thermal technologies, biological technologies could provide a major breakthrough with new generations of firms using 100 percent biodegradable feedstock and advanced biotechnologies.[57]

Biotech firms are beginning to use patented microbes to convert carbon-rich waste into biofuel by a gas fermentation technology or by turning low-grade cooking oils into biodiesel.[58]

An anaerobic digester uses a series of biological processes in which microorganisms break down biodegradable material into a biogas, which is combusted to generate electricity, heat, or vehicle fuel.[59]

WASTE: AN OPEN PROBLEM

And with over 90 percent of waste openly dumped or burned in low-income countries, it is the poor and most vulnerable that are disproportionately affected.

"Poorly managed waste is contaminating the world's oceans, clogging drains and causing flooding, transmitting diseases, increasing respiratory problems from burning, harming animals that consume waste unknowingly, and affecting economic development, such as through tourism," said Sameh Wahba, World Bank Director for Urban and Territorial Development, Disaster Risk Management and Resilience.[60]

14

CAUTIONARY COMMENTS ON CLIMATE CHANGE

How to Cope

It is your human environment that makes climate.

—Mark Twain

In the future, unlike on *Star Trek*, we probably won't be able to control the weather. But instead of just talking about it, we may be able to mitigate, learn to live with, understand, and deal with it.

WHITHER THE WEATHER

Most of the death and destruction from water-bearing storms comes from storm surges caused by winds pushing relentlessly onto the shore. They are more like rollers of high water that form as the windborne ocean crashes inland. They may occur in addition to high tides. For cities that are at increased risk of flooding, architects are moving away from traditional defenses such as dykes, levees, and dams that are expensive to build and maintain and can be overwhelmed, causing massive flooding in extreme conditions. Instead, there are new plans to make space for water in the urban fabric so that life can (mostly) continue with far less damage due to flooding.[1]

Water plazas holding storm rain during floods will act as reservoirs until the water can drain away via naturally permeable pavement and can also be directed into huge underground cisterns or into some build-ings by hydraulic diversion.[2]

For residents of many coastal cities, the future of continual flooding is already here in the form of rising sea levels and frequent, destructive floods. And the problem is only going to get worse.[3]

More than 90 US communities already face chronic inundation from rising seas caused by climate change, and the number could jump to nearly 170 communities in less than twenty years and as many as 670 by the end of the century, according to a study by analysts at the Union of Concerned Scientists.[4] Environmental scientists and engineers are de-vising a range of ways to prevent coastal flooding by diversion, saturat-ing permeable materials, using natural green defenses, limiting erosion, and deflecting wave energy.[5]

Skyscrapers may not suffer the same flooding conditions of smaller buildings—inhabitants can always move up and use the basements ar-eas for drainage. In the future, buildings will be far more self-contained and efficient, but there will always be the problem of isolation when the streets and underground areas are flooded—in fact, that's one reason that future buildings will be more self-sufficient.

HARDCORE CURES

Seawalls and bulkheads (vertical walls that retain soil) and revetments (sloping structures on banks and cliffs) have long been the go-to de-fenses against coastal flooding. Fourteen percent of all continental US shorelines have been armored with these "hard" structures, and it's estimated that nearly one-third of US coastlines will be reinforced by 2100.[6]

But there's a problem: instead of damping wave energy, these ce-ment structures simply deflect wave power to areas "next door." So if waves batter a seawall along one location, their energy will be redi-rected to neighboring properties. That means some areas will experi-ence wave energies and destruction even greater than would be experi-enced if the seawalls weren't even there.[7]

SOFTCORE SOLUTIONS

A more workable and more natural solution is to create "living shore-lines" with various soft, green shore-protecting techniques and technol-ogies that primarily involve natural materials.

Nature has protected potentially flooding lands in a primal way by water-absorbing salt marshes and using shoreline rock structures, oyster beds, coral reefs, and other natural breakwaters to deflect water and protect the shoreline. Now we are striving to re-create what nature created in the beginning, to disperse the terrific force of the oceans. Stabilization projects involve what coastal engineers refer to as shore protection in conjunction with a beach restoration and maintenance plan. Shore protection is generally categorized as "soft" or "hard" solutions. Soft solutions include beach grass plantings, sand/snow drift fencing, and fiber roll technology. A fiber roll is a collection of coconut fibers rolled together, encased in organic netting, and anchored at the base of the coastal slope. And according to the National Oceanic and Atmospheric Administration (NOAA), just fifteen horizontal feet of marshy terrain can absorb 50 percent of incoming wave energy.[8] The growth of oyster reefs has a good chance of protecting a coastal area.

In one project, coastal scientists and volunteers built and installed square wire cages through which marsh plants would grow, creating a slope of plant matter to lessen wave impact, especially when utilized against a wall.[9]

Research suggests that marshes are significantly better than bulkheads at protecting shorelines. In a survey of three coastal regions of North Carolina, it was observed that Hurricane Irene damaged 76 percent of bulkheads. But the shorelines protected by natural marshes sustained little or no damage.[10]

Green growth also provides protection against floods and rising waters. Plant infrastructure can help mitigate the flooding by rainwater that rises when the ground is covered by impermeable surfaces, such as asphalt, pavement, and cement.

Another idea is basically a human-made "bowl" one hundred to three hundred square feet in diameter that is filled with soil, clay, sand, plants, and mulch. These bowls collect stormwater runoff from houses or small buildings so that it can be absorbed by plants or returned to the atmosphere as water vapor.[11]

Some communities have created larger versions of rain gardens known as bioretention systems, created by punching a hole through a clay layer in the ground to increase water draining.[12] Another solution is called a bioswale—that is, a sloped landscape that channels water into vegetation-filled ditches, complementing other green infrastructures.[13]

Green roofs and green walls with thick vegetation can reduce the volume of stormwater runoff from buildings.[14] On hard, impervious land, ground covers can be replaced with a porous surface concrete product called Topmix Permeable that can absorb water at a rate of about 880 gallons per minute, preventing pooling in parking lots and road surfaces.[15]

Living shorelines aren't possible for all places. Take New York City, for example. To combat a powerful storm like Hurricane Sandy would have required establishing many miles of natural habitat, which could not have been possible in an urban waterfront setting.[16]

One inevitable problem is that some cities that suffer habitual flooding are at the lowest and flattest parts of the coastal United States, where the land is also sinking. Climate change will only worsen the plight of these communities and similar ones in other parts of the country.[17]

If global sea levels were to rise just one foot within the next twenty years (a good but gloomy guess), more than one hundred coastal communities would see up to 25 percent of their livable land flooded. A far more pessimistic guesstimate is that by 2100, close to 670 communities would be chronically flooded, including Boston, Newark (New Jersey), Fort Lauderdale, Los Angeles, and all but one of New York City's five boroughs.[18] The latest research suggests that by 2100, up to 60 percent of oceanfront communities on the East and Gulf Coasts of the United States may experience chronic flooding from climate change.[19]

On the West Coast, dire predictions include losing expensive beach front real estate in southern California, airports in Oakland and San Francisco along with forty-two thousand homes along the coast and the inland Sacramento delta.[20]

The final, dismal solution to flooding is simple: pull up stakes and move away.

CREEPING CLIMATE CHANGE

Ultimately, creeping climate change is going to be inherited by future generations. But for residents of many coastal cities, the future is right around the corner—or already peeking at us—in the form of rising sea levels and frequent, destructive floods. Worst-case scenario is that parts of America could experience sea-level rises of as much as eight feet by 2100.[21]

Consequently, this isn't a great time to buy coastal property, especially with a thirty-year mortgage. About 40 percent, or 125 million members, of the US population live in counties located on a coastline.[22] It's not too hard to guess what will happen to property values along the coast as rising ocean levels and extreme climate events put more and more homes in danger of being flooded, swept out to sea in high tide, or destroyed by storm surges or hurricane events. Even in nonextreme circumstances there is the danger that they will collapse along eroding shoreline bluffs or regularly become flooded because they're below the (new) mean high-water line.

Just north of San Diego, the small, coastal town of Del Mar (ironically means "of the sea" in Spanish) is a wealthy enclave full of expensive homes built almost on top of Pacific Ocean bluffs and shoreline. The state of California requires Del Mar and all other coastal communities to create plans for managing their properties in the face of the expected rise of three to six feet in average sea level over the next several decades and to also understand that the land—including cliffs and beaches—will inevitably be taken by the Pacific Ocean.[23]

That actually happened way, way farther inland, in Grand Forks, North Dakota, where the Red River area was inundated so badly that people just abandoned their land and homes and simply moved to higher ground.[24]

Jerry Yudelson, a green guru and author of *Reinventing Green Building,* has stated, "My prediction: this same debate and response is going to happen in city after city as people begin to face up to the reality that 'mitigation' because of climate change damages might very well mean 'abandonment,' of settled areas, leaving them to face the elements."[25]

NASA, the world's leading climate research agency, says all ten of the planet's warmest years since records have been kept occurred in the past twelve years. In some places global sea level rose about seventeen centimeters (6.7 inches) in the last century. In some places the rate in the last decade is nearly double that.[26] All three major global surface temperature reconstructions show that Earth has warmed since 1880.[27]

The growing heat island effect in cities, which contributes to global warming, can be partially handled by using green building materials that don't capture heat.[28] Dark-colored asphalt absorbs between 80 and 95 percent of the sun's rays, heating up not just the streets themselves but the entire surrounding area. And according to the Bureau of Street Services, the LA streets that have been rendered lighter in color with a grayish-white coating known as CoolSeal renders the streets ten to fifteen degrees cooler on average.[29] Using green building materials and alternatives could be one of the keys to saving our coastal communities.

Protecting places like New York City by dissipating big waves is not practical. It makes more sense to create larger versions of rain gardens known as bioretention gardens, a former wetland area punched with holes through the clay layer to increase infiltration. The garden fills with rain and floodwater, turning it into a full wetland that serves as habitat for wildlife.[30]

Green roofs and green walls blanketed with vegetation can reduce the volume of storm water runoff from buildings. New York has been weighing very ambitious plans for defending itself against further assaults from the sea by constructing a combination of a large chain of artificial islands and a giant floodwall.[31]

Perhaps more practical plans, as reported by the *New York Times*, includes the idea that the risk of future flooding is changing the way that buildings are designed in the city. Top-floor penthouses might be replaced with emergency generators that can provide enough power for residents to remain in their apartment for as long as a week, and ground floors are being built with materials that can tolerate floods.[32]

SHAKE, RATTLE, AND ROLL

The occurrence of earthquake tremblers will become more problematic as more people move to urban environments and as structures grow. Luckily, over the last few decades, architects and engineers have devised a number of clever technologies to ensure that houses, multidwelling units, and skyscrapers bend but don't break. As a result, after the building shakes and suffers only minimal damage, inhabitants can walk out unharmed.[33]

Earthquakes also produce land and surface waves. The former travel rapidly through the earth's interior. The latter travel more slowly through the upper crust and include a subset of waves—known as Rayleigh waves—that move the ground vertically. This up-and-down motion causes most of the shaking and damage associated with an earthquake.[34]

Devices such as isolation systems and dampers are designed to reduce vibrations and, as a result, the damage of structures. These devices include a novel barrier that absorbs an earthquake's ground waves. The "ViBa" is essentially a box containing a solid central mass connected to the foundations of buildings through the soil. It is held in place by springs, allowing the mass to move back and forth and absorb the vibrations created by seismic waves. It should be able to absorb a significant portion of that energy, with a subsequent 40 to 80 percent reduction of seismic response.[35]

Another system involves "floating" a building above its foundation on lead and rubber bearings, which contain a solid lead core wrapped in alternating layers of rubber and steel. Steel plates attach the bearings to the building foundation and, when an earthquake hits, allow the foundation to move without moving the structure above it.[36]

Yet another system uses a cushion of forced air. When sensors on the building detect seismic activity, a network of sensors communicates with an air compressor that, within a half-second of being alerted, forces air between the building and its foundation. The cushion of air lifts the structure up to 1.18 inches off the ground, isolating it from the quake force. When the earthquake subsides, the compressor turns off, and the building settles back down to its foundation.[37]

Shock absorbers in a building slow down and reduce the magnitude of vibrations by turning the kinetic energy of the suspension devices into heat energy dissipated through hydraulic fluid, a process known as damping. Dampers can be placed on each building level with one end attached to a column and the other end attached to a beam. Each damper with a piston head moves inside a cylinder filled with silicone oil. When an earthquake strikes, the horizontal motion of the building causes the piston in each damper to push against the oil, damping the motion.[38]

Another solution, especially for skyscrapers, involves suspending an enormous mass near the top of the structure. Steel cables support the mass, while viscous fluid dampers lie between the mass and the building it's trying to protect. When seismic activity causes the building to sway, the mass moves in the opposite direction, dissipating the energy.[39]

Another idea is a controlled rocking system in which the steel frames that make up the structure are elastic and allowed to rock on top of the foundation, which means they can pull the entire structure upright when the shaking stops. The final components are the replaceable steel fuses placed between two frames or at the bases of columns. The metal teeth of the fuses absorb seismic energy as the building rocks. If they "blow" during an earthquake, they can be replaced relatively quickly and cost-effectively to restore the building to its original, ribbon-cutting form.[40]

One more design is a rocking core-wall at the ground level that prevents the concrete in the wall from being permanently deformed. To accomplish this, engineers reinforce the lower two levels of a building with steel, and they incorporate post-tensioning forms along the entire height. In post-tensioning systems, steel tendons are threaded through the core wall. The tendons act like rubber bands, which can be tightly stretched by hydraulic jacks to increase the tensile strength of the core-wall.[41]

Scientists call another possible system a "seismic invisibility cloak" for its ability to render a building invisible to surface waves. Engineers believe they can fashion a "cloak" out of one hundred concentric plastic rings buried beneath the foundation of a building. As seismic waves approach, they enter the rings at one end and become contained within the system. Harnessed within the "cloak," the waves can't impart their energy to the structure above. They simply pass around the building's foundation and emerge on the other side, where they exit the rings.[42]

A "shape memory" alloy can endure heavy strains and still return to its original shape. Engineers are experimenting with these so-called smart materials as replacements for traditional steel-and-concrete construction. One promising alloy is nickel titanium, or nitinol, which offers 10 to 30 percent more elasticity than steel.[43] Researchers compared the seismic performance of bridge columns made of steel and concrete with columns made of nitinol and concrete. The columns made of nitinol and concrete allow for shape memory, which means the structures can endure heavy strains and still return to their original shape. They far outperformed the traditional materials on all levels and experienced far less damage.[44]

Another promising solution, much easier to implement, requires a technology known as fiber-reinforced plastic wrap. Manufacturers produce these wraps by mixing carbon fibers with binding polymers, such as epoxy polyester, vinyl ester, or nylon, to create a lightweight but incredibly strong composite material.

Engineers simply wrap the material around concrete support columns of bridges or buildings and then pump pressurized epoxy into the gap between the column and the material, offering significantly higher strength and ductility (strength before rupture). Earthquake-damaged columns can be repaired with carbon-fiber wraps and can be 24 to 38 percent stronger than unwrapped columns.[45]

Some marine biologists have suggested using the sticky fibers of sea mussels, known as byssal threads, because the flexible strands absorb the shock and dissipate the energy of the ocean. Researchers have even calculated the exact ratio of stiff-to-flexible fibers—80:20—that gives the mussel its stickiness, and they are developing materials that mimic the mussel's strength.[46]

Another biomimic is spider silk. It's stronger than steel pound for pound (just ask Peter Parker), and its dynamic response under heavy strain makes it unique. When the silk is tugged, the threads are initially stiff, then stretchy, then stiff again. It's this response that makes spider webs so resilient and spider thread such a tantalizing material to mimic in the next generation of earthquake-resistant construction.[47] Scientists are presently trying to solve the problem of mass amounts of the spidey silk by hybridizing the genes of a silk spider with a goat. The silk is part of the goat's milk and is purified producing the spider silk protein into much, higher quantities.[48]

Another possibility is for humble cardboard tubes to be coated with polyurethane and combined with wood as primary framing elements. Because the cardboard-and-wood structure is extremely light and flexible, it performs much better than concrete during seismic events. If it collapses, it's far less likely to crush people gathered inside.[49] And it's easy to reproduce.

For something more personal, how about Capsule K107, an egg-shaped pod? Once a quake hits, you step inside, close the door, and hide inside and ride out an earthquake. The survival capsule is a spherical, reinforced metal ball that can withstand being crushed and will also float in the case of a tsunami—and it can be lived in for up to a month. A basic version goes for $2,400, but a top-of-the-line model costs $10,000. The pods feature a pouch for human waste, an air purifier, and a vapor condenser to supply drinking water.[50]

WEATHER WAY AWAY FROM HOME

The moon has little to no atmosphere, which means there's no wind and very little weather. The surface can reach temperatures of up to 253°F during the day and drop to –243° at night.[51]

On our neighboring planets there's going to be very little enjoyment, if any, of extraterrestrial strolls, rainstorms, or snowfalls. And as for enjoying a summer breeze, winds in the strongest Martian storms top out at about 60 mph, about three-quarters the speed of a Category 1 hurricane on Earth.[52]

Because of its thin atmosphere and its greater distance from the sun, the surface temperature of Mars is much colder than that of Earth. The average temperature on Mars ranges from a balmy 70 to a brisk –225°F.[53] The atmosphere of Mars is also roughly one hundred times thinner than Earth's, but it is still thick enough to support some weather phenomena, like clouds and winds.[54] Mars is infamous for intense dust storms that sometimes kick up enough dust to be seen by telescopes on Earth. Every year there are some big dust storms that can completely cover the planet and block out the sun.[55]

Because Venus is closer to the sun, and because it has a thick atmosphere of heat that traps carbon dioxide and sulfuric acid, the average temperature is about 860°. On its surface is a crushing atmosphere ninety-three times heavier than on Earth, plus a thick, sulfuric acid–laced atmosphere in which there are lightning storms very much like those on Earth.[56] In fact, scientists claim that the only way to colonize the planet is to live high above the hostile atmosphere in giant dirigibles capable of housing thousands of "cloud people."[57]

CLIMATE CHANGE CONJECTURE

What you see when you look out the window on a daily basis is weather. Climate change refers to alterations in the atmosphere such as temperature, precipitation, ice melt, the jet stream, and wind patterns, measured over hundreds, even thousands, of years. The problem is that there is evidence that temperatures within human history have never increased as rapidly as in the past one hundred years. According to scientists at NASA's Goddard Institute for Space Studies (GISS), globally, the warmest temperatures were those of 2015, 2016, 2017, and made 2018 the second-warmest year on record behind only 2017.[58] That increase is driven largely by human activities, mainly the use of fossil fuels.[59]

An international team of researchers formed some conclusions by running computer simulations and predicted what future weather patterns around the globe would look like if levels of greenhouse gases continued to rise as expected.

The simulations were run three times, with greenhouse gases set at either low, medium, or high. All three scenarios predicted increases in extreme weather conditions but differed regarding their frequency. Here is what they figure will occur by the end of the century:

- Most areas above latitude 40° north, including parts of Canada and northern parts of the United States will experience more days of heavy rain, defined as more than 0.40 inches.
- Dry spells, which can lead to drought, could lengthen significantly across the western United States.
- On the bright side, the average growing season could increase significantly across most of North America.[60]

WHETHER THE WEATHER CAN BE CHANGED

There are many factors that influence global weather, but one key message is that climate change and warming are likely to make extreme weather on both ends of the spectrum more common.

With the threat of rising global temperatures, increased rainfall, and severe droughts, and more and larger events, scientists are racing to develop technologies that will actually change the weather, like "making it rain" to mitigate the severe droughts experienced especially in the Southwest. One such older technology is cloud seeding; it is the process of spreading either dry ice, or more commonly, silver iodide aerosols, into the upper part of clouds to try to stimulate the precipitation process and form rain. Silver iodide gets sprinkled into clouds by airplanes or blasted up into clouds on rockets, or winds are used to naturally transport the silver iodide into the clouds. The chemical has a very similar structure to ice, so it will bond to clouds, making them increasingly heavy until they let loose their moisture.[61]

Some of these technologies seem promising, but there's no telling about consequences when we start messing with Mother Nature.

CO$_2$ AND THE CHANGE IN CLIMATE

Carbon dioxide is not a total bogeyman: it retains heat in the atmosphere and keeps the planet warm enough to sustain life. But it is a problem when we end up with too much, especially through the burning of fossil fuels. Certain gases in the atmosphere block heat from escaping. Long-lived gases that remain semipermanently in the atmosphere and do not respond physically or chemically to changes in temperature are described as "forcing" climate change. Gases, such as water vapor, which respond physically or chemically to changes in temperature are seen as "feedbacks," causing a rise in temperature.[62]

The four most common greenhouse gases are carbon dioxide (CO_2), methane (CH_4), nitrous oxide (N_2O), and water vapor (H_2O).[63] Impacts of climate change include:

- Sea level rises due to the melting of ice sheets; sea levels in the United States could rise more than twenty inches in the twenty-first century
- Extinction of animals migrating to higher elevations or away from the equator as temperatures warm, and the extinction of many forms of plant life
- Absorption by the oceans of extra carbon dioxide in the atmosphere, making it difficult for corals and microorganisms to survive and disrupting the food supply for other sea animals
- Increased fires in forests and brush, exacerbating forest decline, drought, and devastating wildfires
- Stressed water supplies due to increased drought.

Whether you choose to bury your head in the oncoming rising tide or, instead, pay attention to people who have made it their life's work to follow the climate of our earth, it seems that erring on the side of reason and safety makes far more sense than relying on ignorance and hearsay. Roll the dice offered by some "leaders" and politicians if you choose; just don't whine when you get your feet wet or your crops fail.

NOTES

1. SEEKING SUPERCITIES

1. William E. Rees, "Building More Sustainable Cities," *Scientific American*, March 1, 2009, https://www.scientificamerican.com/article/building-more-sustainable-cities.

2. Jeff Desjardins, "Animation: The World's Largest Megacities by 2100," Visual Capitalist, July 16, 2018, https://www.visualcapitalist.com/worlds-20-largest-megacities-2100.

3. "America Moves to the City," Khan Academy, https://www.khanacademy.org/humanities/us-history/the-gilded-age/gilded-age/a/america-moves-to-the-city.

4. Wikipedia, s.v. "1860 United States Census," last modified March 8, 2019, 2:22, https://en.wikipedia.org/wiki/1860_United_States_Census.

5. "Chapter 25: America Moves to the City, 1865–1900," CourseNotes, https://course-notes.org/us_history/notes/the_american_pageant_14th_edition_textbook_notes/chapter_25_america_moves_to_the_ci.

6. Wikipedia, s. v. "Urbanization in the United States," last modified July 25, 2019, 13:58, https://en.wikipedia.org/wiki/Urbanization_in_the_United_States.

7. "Rise of Industrial America, 1876–1900: City Life in the Late 19th Century," American Memory Timeline, Library of Congress, http://www.loc.gov/teachers/classroommaterials/presentationsandactivities/presentations/timeline/riseind/city.

8. Matt Ridley, *The Rational Optimist* (New York: HarperCollins, 2010).

9. SeniorLiving.org, "1800–1990: Changes in Urban/Rural U.S. Population," https://www.seniorliving.org/history/1800-1990-changes-urbanrural-us-population.

10. Med Amine Bensefia and Abdelhafidh Benmansour, "The Ambiguity of the American Dream and the Shift to Hollywood Dream" (master's thesis,

Abou Bakr Belkaild University, Tlemcen, 2014–2015), http://dspace.univ-tlemcen.dz/bitstream/112/8031/1/amine-bensefia.pdf.

11. Wikipedia, s.v. "Mortgage Discrimination," last modified August 1, 2019, 22:42, https://en.wikipedia.org/wiki/Mortgage_discrimination.

12. Nikole Hannah-Jones, "Living Apart: How the Government Betrayed a Landmark Civil Rights Law," ProPublica, last modified July 8, 2015, https://www.propublica.org/article/living-apart-how-the-government-betrayed-a-landmark-civil-rights-law.

13. Rachel Becker, "World Population Expected to Reach 9.7 Billion by 2050," *National Geographic*, July 31, 2015, https://news.nationalgeographic.com/2015/07/world-population-expected-to-reach-9-7-billion-by-2050.

14. John Vidal, "The 100 Million City: Is 21st Century Urbanisation Out of Control?" *The Guardian*, last modified April 13, 2018, https://www.theguardian.com/cities/2018/mar/19/urban-explosion-kinshasa-el-alto-growth-mexico-city-bangalore-lagos; Wikipedia. s.v. "Projections of Population Growth," last modified August 16, 2019, 17:11, https://en.wikipedia.org/wiki/Projections_of_population_growth.

15. Edith M. Lederer, "UN Report: By 2030 Two-Thirds of World Will Live in Cities," AP News, May 18, 2016, https://www.apnews.com/40b530ac84ab4931874e1f7efb4f1a22.

16. "U.S. Cities Are Home to 62.7 Percent of the U.S. Population, but Comprise Just 3.5 Percent of Land Area," US Census Bureau, March 4, 2015, https://www.census.gov/newsroom/press-releases/2015/cb15-33.html.

17. Nate Berg, "U.S. Urban Population Is Up . . . but What Does 'Urban' Really Mean?" CityLab, March 26, 2012, https://www.citylab.com/equity/2012/03/us-urban-population-what-does-urban-really-mean/1589.

18. Mark Swilling, "The Curse of Urban Sprawl: How Cities Grow, and Why This Has to Change," *The Guardian*, July 12, 2016, https://www.theguardian.com/cities/2016/jul/12/urban-sprawl-how-cities-grow-change-sustainability-urban-age.

19. Deirdre Pfeiffer, Genevieve Pearthree, and Meagan Ehlenz, "Efforts to Attract Millennials Are Reshaping Downtown Areas," *Government Technology*, January 18, 2018, https://www.govtech.com/fs/infrastructure/Efforts-to-Attract-Millennials-Are-Reshaping-Downtown-Areas.html.

20. Lyman Stone, "American Women Are Having Fewer Children Than They'd Like," *New York Times*, February 13, 2018, https://www.nytimes.com/2018/02/13/upshot/american-fertility-is-falling-short-of-what-women-want.html.

21. Sommer Mathis, "Of Course the Suburbs Aren't Dying—They're Not All the Same," CityLab, January 23, 2015, https://www.citylab.com/equity/2015/01/of-course-the-suburbs-arent-dying-theyre-not-all-the-same/384781.

22. Mathis, "Of Course the Suburbs Aren't Dying."

23. Madeline Stone, "Why McMansions Were Doomed Investments from the Start," *Business Insider*, September 10, 2016, https://www.businessinsider.com/why-is-it-called-a-mcmansion-2016-9.

24. New Hampshire Public Radio, "Are American Suburbs Dying?" *Here and Now*, March 7, 2017, https://www.nhpr.org/post/are-american-suburbs-dying.

25. Rent Editorial Team, "Millennial Generation Choosing to Rent," *Rent* (blog), May 14, 2015, https://www.rent.com/blog/millennial-generation-renting.

26. Neale Godfrey, "The Young and the Restless: Millennials on the Move," *Forbes*, October 2, 2016, https://www.forbes.com/sites/nealegodfrey/2016/10/02/the-young-and-the-restless-millennials-on-the-move.

27. Megan Gorman, "The Conversation Gen Xers Must Have with Their Boomer Parents," *Forbes*, April 30, 2019, https://www.forbes.com/sites/megangorman/2019/04/30/the-conversation-gen-xers-must-have-with-their-boomer-parents.

28. Denise Mann, "6 Signs Your Commute Is Making You Sick—And What to Do About It," *Reader's Digest*, https://www.rd.com/health/conditions/commute-making-you-sick.

29. Marlynn Wei, "Commuting: 'The Stress That Doesn't Pay,'" *Psychology Today*, January 12, 2015, https://www.psychologytoday.com/us/blog/urban-survival/201501/commuting-the-stress-doesnt-pay.

30. Annette Schaefer, "Commuting Takes Its Toll," *Scientific American*, October 1, 2005, https://www.scientificamerican.com/article/commuting-takes-its-toll.

31. Matt Powell, "Sneakernomics: How Golf Lost the Millennials," *NPD* (blog), February 23, 2017, https://www.npd.com/wps/portal/npd/us/blog/2017/sneakernomics-how-golf-lost-the-millennials.

32. Emily Badger, "Quantifying the Cost of Sprawl," CityLab, May 21, 2013, https://www.citylab.com/equity/2013/05/quantifying-cost-sprawl/5664.

33. "Atlanta BeltLine Overview: The Atlanta Beltline in 5," Atlanta Belt-Line, https://beltline.org/about/the-atlanta-beltline-project/atlanta-beltline-overview.

34. United Nations, "World's Population Increasingly Urban with More Than Half Living in Urban Areas," July 10, 2014, https://www.un.org/en/development/desa/news/population/world-urbanization-prospects-2014.html.

35. Ondel Hylton, "432 Park in Numbers: New Renderings and Superlatives Will Blow You Away," 6SQFT, November 11, 2015, https://www.6sqft.com/432-park-in-numbers-new-renderings-and-superlatives-will-blow-you-away.

36. Jonathan O'Callaghan, "Welcome to the Cities of the Future: 'Impossible Engineering' Predicts Cows on Skyscrapers, 3D Printed Homes and Underwater Arenas in the Next 100 Years," *Daily Mail Online*, May 25, 2015, https://www.dailymail.co.uk/sciencetech/article-3096283/Welcome-cities-future-Impossible-Engineering-predicts-cows-skyscrapers-3D-printed-homes-underwater-arenas-100-years.html.

37. Benjamin Preston, "America's Infrastructure Still Rates No Better Than D+, Engineering Experts Say," *The Guardian*, March 9, 2017, https://www.

theguardian.com/us-news/2017/mar/09/america-infrastructure-rating-problems-engineers-report.

38. Cadie Thompson, "There's a $1 Trillion Crisis Threatening the American Way of Life as We Know It," *Business Insider*, March 6, 2017, https://www.businessinsider.com/american-infrastructure-falling-apart-2017-2.

39. Cynthia Drescher, "The 10 Fastest Trains in the World," *Conde Nast Traveler*, March 27, 2018, https://www.cntraveler.com/stories/2016-05-18/the-10-fastest-trains-in-the-world.

40. Enrique Penalosa, "This Is What the Cities of the Future Will Look Like," HuffPost, May 8, 2016, https://www.huffingtonpost.com/enrique-penalosa/cities-future_b_7216732.html.

41. Penalosa, "This Is What the Cities of the Future Will Look Like."

42. Kaid Benfield, "Pedestrian Perfection: The 11 Most Walk-Friendly U.S. Cities," May 4, 2011, *The Atlantic*, https://www.theatlantic.com/national/archive/2011/05/pedestrian-perfection-the-11-most-walk-friendly-us-cities/238337.

43. O'Callaghan, "Welcome to the Cities of the Future."

44. "How the World Will Look in the Future before Apocalypse," *Religion* (blog), November 5, 2013, https://religionchristian.blogspot.com/2013/11/how-world-will-look-in-future-before.html.

45. Laura Schier, "Queens Experiences Highest Rent Increase in the Country," *Elegran* (blog), August 7, 2018, https://www.elegran.com/blog/2018/08/queens-experiences-highest-rent-increase-in-the-country.

46. Don Willmott, "Building the World's First Carbon-Neutral City," *Smithsonian*, September 22, 2014, https://www.smithsonianmag.com/innovation/building-the-worlds.

47. Alan Davis, "Do You Have Any Idea How Much Green Spaces in Cities Are Actually Worth?" AlterNet, February 27, 2018, https://www.alternet.org/2018/02/how-valuable-are-green-spaces-our-cities.

48. "Get Connected: Smart Buildings and the Internet of Things," *Facility Executive*, May 28, 2019, https://facilityexecutive.com/2019/05/get-connected-smart-buildings-iot-buildings-management.

49. Mark J. Perry, "Fossil Fuels Will Continue to Supply >80% of US Energy through 2040, While Renewables Will Play Only a Minor Role," *Carpe Diem* (blog), AEI, December 16, 2013, https://www.aei.org/publication/fossil-fuels-will-continue-to-supply-80-of-us-energy-through-2040-while-renewables-will-play-only-a-minor-role.

50. Bill Chappell, "California Gives Final OK to Require Solar Panels on New Houses," NPR, December 6, 2018 , https://www.npr.org/2018/12/06/674075032/california-gives-final-ok-to-requiring-solar-panels-on-new-houses.

51. Michael J. Coren, "There Is a Point at Which It Will Make Economic Sense to Defect from the Electrical Grid," *Quartz*, June 29, 2017, https://qz.com/1017457/there-is-a-point-at-which-it-will-make-economic-sense-to-defect-from-the-electrical-grid.

52. Paul Marks, "Vertical Farms Sprouting All over the World," *New Scientist*, January 15, 2014, https://www.newscientist.com/article/mg22129524-100-vertical-farms-sprouting-all-over-the-world.

53. "Detroit: A City in Decline—in Pictures," *The Guardian*, July 19, 2013, https://www.theguardian.com/world/gallery/2013/jul/19/detroit-goes-bankrupt-in-pictures.

54. Ann O'M. Bowman and Michael A. Pagano, "Vacant Land in Cities: An Urban Resource," Brookings, January 1, 2001, https://www.brookings.edu/research/vacant-land-in-cities-an-urban-resource.

55. "Vacant and Abandoned Properties: Turning Liabilities into Assets," *Evidence Matters* (Winter 2014), https://www.huduser.gov/portal/periodicals/em/winter14/highlight1.html.

56. "Vacant and Abandoned Properties."

57. Bowman and Pagano, "Vacant Land in Cities."

58. "Federal Grant Opportunities," Reconnecting America, http://reconnectingamerica.org/resource-center/federal-grant-opportunities.

59. Will Worley, "Climate Change: Flooding Caused by Global Warming to Put One Billion at Risk by 2060, Charity Warns," *The Independent*, May 16, 2016, https://www.independent.co.uk/news/world/climate-change-related-flooding-one-billion-people-risk-2060-warns-charity-a7031241.html.

60. "Sci-Fi Cities & Mega Cities of the Future, as Imagined or Not?" Citi IO, May 30, 2016, https://citi.io/2016/05/30/sci-fi-cities-mega-cities-of-the-future-as-imagined-or-not.

2. HABITATS FOR INHABITANTS

1. Brian Wang, "World Wealth, People and Cities in 2050–2060," *Nextbigfuture* (blog), September 14, 2018, https://www.nextbigfuture.com/2018/09/world-wealth-people-and-cities-in-2050-2060.html.

2. Bret Boyd, "Megacities and Complexity," Grayline, https://graylinegroup.com/megacities-and-complexity.

3. Liam Tung, "Google's Massive Expansion Plan: Its Own Village with Up to 8,000 Homes," ZDnet, December 10, 2018, https://www.zdnet.com/article/googles-massive-expansion-plan-its-own-village-with-up-to-8000-homes.

4. Jennifer Elias, "Google to Invest $1 Billion in San Francisco Bay Area Housing amid Regional Expansion," CNBC, June 18, 2019, https://www.cnbc.com/2019/06/18/google-to-invest-1-billion-in-san-francisco-bay-area-housing.html.

5. Katie Avis-Riordan, "Top 10 Bathroom Innovations People Would Love to Have in the Future," *House Beautiful*, March 16, 2018, https://www.housebeautiful.com/uk/lifestyle/a19432401/bathroom-technology-innovations-future.

6. Kate Baggaley, "Here's How Smart Toilets of the Future Could Protect Your Health," NBC News, January 23, 2019, https://www.nbcnews.com/mach/

science/here-s-how-smart-toilets-future-could-protect-your-health-ncna961656.

7. Juhan Sonin, "From Bathroom to Healthroom," GoInvo, https://www.goinvo.com/features/from-bathroom-to-healthroom.

8. "The Bluetooth Connected Toothbrush That Reveals Everything," Philips/Sonicare, https://www.usa.philips.com/c-m-pe/dental-professionals/products/tooth-brushes/flexcare-platinum-connected#triggername=feature_smart-sensor.

9. Gabrielle Golenda, "High Tech Tuchus: 6 Smart Toilets," *Architect's Newspaper*, February 12, 2018, https://archpaper.com/2018/02/high-tech-tuchus-6-smart-toilets.

10. Michelle Ullman, Lulu Chang, and Steven John, "The Best Air Purifier You Can Buy," *Business Insider*, November 29, 2018, https://www.businessinsider.com/best-air-purifier.

11. "Causes of Obstructive Sleep Apnea," WebMD, August 24, 2017, https://www.webmd.com/sleep-disorders/sleep-apnea/obstructive-sleep-apnea-causes#1.

12. "Digital Parenting: The Best Baby Tech and Connected Baby Monitors," Wareable, June 12, 2017, https://www.wareable.com/health-and-wellbeing/best-smart-baby-monitors.

13. "The Only Mobile Medical Alert with No Monthly Monitoring Fees," LiveLife, https://livelifealarms.com/product/mobile-medical-alert.

14. Lauren Silverman, "'Smart' Pill Bottles Aren't Always Enough to Help the Medicine Go Down," Shots—Health News, August 22, 2017, https://www.npr.org/sections/health-shots/2017/08/22/538153337/smart-pill-bottles-arent-enough-to-help-the-medicine-go-down.

15. Ipke Wachsmuth, "Robots Like Me: Challenges and Ethical Issues in Aged Care," *Frontiers in Psychology*, April 3, 2018, https://www.ncbi.nlm.nih.gov/pmc/articles/PMC5892289.

16. Renée Lynn Midrack, "What Is a Smart Refrigerator?" Lifewire, updated May 16, 2019, https://www.lifewire.com/smart-refrigerator-4158327.

17. "Samsung TVs Use Your Wall as A Screen Saver to Blend In," Cnet, March 8, 2018, https://www.cnet.com/videos/samsung-tvs-use-your-wall-as-a-screen-saver-to-blend-in.

18. Vinay Prajapati, "The Future of Smart TVs," *TechPrevue* (blog), April 29, 2018, https://www.techprevue.com/future-smart-tvs-detail.

19. "Smart Smoke Detectors Are the Future of Fire Protection," *Smart Home Blog*, Go Konnect, November 11, 2017, https://www.gokonnect.ie/blog/smart-smoke-detectors-are-the-future-of-fire-protection.

20. "Ten Ways to Reduce Greenhouse Gases," Town of East Gwillimbury, http://www.eastgwillimbury.ca/Services/Environment/Ten_Ways_to_Reduce_Greenhouse_Gases.htm

21. Raphael Slade and Ausilio Bauen, "Micro-Algae Cultivation for Biofuels: Cost, Energy Balance, Environmental Impacts and Future Prospects," *Biomass and Bioenergy* 53 (June 2013): 29–38, https://doi.org/10.1016/j.biombioe.2012.12.019.

22. Veolia Environment and the London School of Economics, "Nanoscopic Robot Recyclers: The Future of Waste Management," *Waste Management World*, November 22, 2013, https://waste-management-world.com/a/nanoscopic-robot-recyclers-the-future-of-waste-management.

23. Ken Lynch, "RFID: Integrating RFID Technology into Waste Management Operations," *Waste Advantage*, August 5, 2014, https://wasteadvantagemag.com/integrating-rfid-technology-into-waste-management-operations.

24. Annie Kane, "Future Could See Waste-Sorting 'Nanobots,'" *Resource*, November 21, 2013, https://resource.co/sustainability/article/future-could-see-waste-sorting-nanobots.

25. Uptin Saiidi, "This New Urban Jungle in Singapore Could Be the Future of Eco-Friendly Buildings," CNBC, August 8, 2018, https://www.cnbc.com/2018/08/08/this-new-urban-jungle-in-singapore-could-be-the-future-of-eco-friendly.html.

26. "About USGBC," https://new.usgbc.org/about.

27. Jill Tunstall, "The Garden of the Future?" *The Guardian*, December 6, 2007, https://www.theguardian.com/environment/2007/dec/06/ethicalliving.conservation.

28. Laurie Goodman, "Manufactured Homes Could Ease the Affordable Housing Crisis. So Why Are So Few Being Made?" *Urban Wire* (blog), Urban Institute, January 29, 2018, https://www.urban.org/urban-wire/manufactured-homes-could-ease-affordable-housing-crisis-so-why-are-so-few-being-made.

29. David Tal, "Forecast: Housing Prices Crash as 3D Printing and Maglevs Revolutionize Construction," Quantumrun, https://www.quantumrun.com/prediction/housing-prices-crash-3d-printing-and-maglevs-revolutionize-construction-future-cities-p3.

30. J. Baldwin, "Dymaxion House," Buckminster Fuller Institute, https://www.bfi.org/about-fuller/big-ideas/dymaxion-world/dymaxion-house.

31. Lucie Gaget, "3D Printing for Construction: What Is Contour Crafting?" Sculpteo, June 27, 2018, https://www.sculpteo.com/blog/2018/06/27/3d-printing-for-construction-what-is-contour-crafting.

32. Mary Hall, "How Does Supply and Demand Affect the Housing Market?" Investopedia, July 6, 2019, https://www.investopedia.com/ask/answers/040215/how-does-law-supply-and-demand-affect-housing-market.asp.

33. Kirsty Sier, "Ecocapsule: A Self-Sustaining Smart House Primed for Off-Grid Living," *Architecture and Design*, June 7, 2017, https://www.architectureanddesign.com.au/news/the-ecocapsule-a-self-sustaining-smart-house-prime#.

34. Stuart Miles, "Putting the Home in HomePod: The New Housing Development That's Smart from Day One," Pocket-Lint, April 1, 2018, https://www.pocket-lint.com/smart-home/news/apple/144048-putting-the-home-in-homepod-the-new-housing-development-that-s-smart-from-day-one.

35. "The Tiny House Movement?" Tiny Life, https://thetinylife.com/what-is-the-tiny-house-movement.

36. "Concrete Pipe House," *Tiny House Blog*, https://tinyhouseblog.com/tiny-house-concept/concrete-pipe-house.

37. Andrea Lo, "Are These Concrete Water Pipes the Answer to Housing Problem?" CNN, January 23, 2018, https://www.cnn.com/style/article/opod-tube-home-hong-kong/index.html.

38. Megan Schires, "IKEA Explores Future Urban Living for the Many," ArchDaily, June 4, 2019, https://www.archdaily.com/918417/ikea-explores-future-urban-living-for-the-many.

39. "Building Blocks: Open Source Spaces for the Many," SPACE10, April 25, 2018, https://space10.io/project/building-blocks.

40. Jenny Xle, "This Aluminum Prefab Was Designed to Be Mobile, Stackable," Curbed, December 16, 2015, https://www.curbed.com/2015/12/16/10620338/prefab-home-mobile-new.

41. "This 'Pop-Up' House Will Make IKEA-Lovers Swoon," *Huffington Post*, March 18, 2014, https://www.huffingtonpost.com/2014/03/18/pop-up-house_n_4988251.html.

42. Matthew Hutson, "Watch This Robot Construct the World's Biggest Botmade Building by Itself," *Science*, April 26, 2017, https://www.sciencemag.org/news/2017/04/watch-robot-construct-world-s-biggest-botmade-building-itself.

43. Catherine Winter, "30 Awesome DIY Projects That You've Never Heard Of," Life Hack, https://www.lifehack.org/articles/lifestyle/30-easy-and-awesome-diy-projects.html.

44. Eoghan Macguire, "The Chinese Firm That Can Build a Skyscraper in a Matter of Weeks," CNN, last modified June 26, 2015, https://www.cnn.com/2015/06/26/asia/china-skyscraper-prefabricated/index.html.

45. Macguire, "The Chinese Firm That Can Build a Skyscraper in a Matter of Weeks."

46. Goodman, "Manufactured Homes Could Ease the Affordable Housing Crisis."

47. Wikipedia, s.v. "Biomimetic Architecture," last modified August 11, 2019, 18:13, https://en.wikipedia.org/wiki/Biomimetic_architecture.

48. Zach Mortice, "Nature Does It Better: Biomimicry in Architecture and Engineering," *Redshift* (blog), Autodesk, July 11, 2016, https://www.autodesk.com/redshift/biomimicry-in-architecture.

49. Ann O'M. Bowman and Michael A. Pagano, "Vacant Land in Cities: An Urban Resource," Brookings Institution, January 1, 2001, https://www.brookings.edu/research/vacant-land-in-cities-an-urban-resource.

50. Bowman and Pagano, "Vacant Land in Cities."

51. Matthew Dolan, "Cost to Clear Detroit's Blight Said to Top $850 Million," *Wall Street Journal*, May 27, 2014, https://www.wsj.com/articles/eliminating-detroits-blighted-buildings-seen-costing-2-billion-1401201530.

52. Patrick Sisson, "The High Cost of Abandoned Property, and How Cities Can Push Back," Curbed, June 1, 2018, https://www.curbed.com/2018/6/1/17419126/blight-land-bank-vacant-property.

53. Bowman and Pagano, "Vacant Land in Cities."

54. Willem Marx, "Michelin-Starred Chef Massimo Bottura Is Reinventing the Soup Kitchen," *Departures*, September 19, 2017, https://www.departures.com/travel/restaurants/massimo-bottura-food-for-soul-project.

55. Richard Paradis, "Retrofitting Existing Buildings to Improve Sustainability and Energy Performance," WBDG, National Institute of Building Sciences, last modified August 15, 2016, https://www.wbdg.org/resources/retrofitting-existing-buildings-improve-sustainability-and-energy-performance.

56. Marisa Kendall, "Life after the Bay Area: Fleeing Residents Feel Heartbreak, Joy," *Mercury News*, last modified July 17, 2018, https://www.mercurynews.com/2018/07/15/life-after-the-bay-area-fleeing-residents-feel-heartbreak-joy.

57. Chava Gourarie, "Rental Market Is at Peak Concession," *Real Deal*, January 11, 2018, https://therealdeal.com/2018/01/11/rental-market-is-at-peak-concessions.

58. Emma Green, "Homelessness Is Up in New York City, but It's Down Everywhere Else," *The Atlantic*, December 13, 2013, https://www.theatlantic.com/business/archive/2013/12/homelessness-is-up-in-new-york-city-but-its-down-everywhere-else/282315.

59. Richard Florida, "Young People's Love of Cities Isn't a Passing Fad," City Lab, May 28, 2019, https://www.citylab.com/life/2019/05/urban-living-housing-choices-millennials-move-to-research/590347.

60. Richard Fry, "Millennials Projected to Overtake Baby Boomers as America's Largest Generation," Pew Research, https://www.pewresearch.org/fact-tank/2018/03/01/millennials-overtake-baby-boomers.

61. Felicia Fuller, "A Comparison of LEED and Green Globes," GBRI, January 28, 2016, https://www.gbrionline.org/a-comparison-of-leed-and-green-globes.

62. Jerry Yudelson, "Green Building Megatrends: Part 2," *Reinventing Green Building* (blog), November 6, 2015, https://www.reinventinggreenbuilding.com/news/2015/11/2/green-building-megatrends-part2.

63. Amy Marpman, "LEED vs. Green Globes," *It's the Environment, Stupid* (blog), August 27, 2008, http://itstheenvironmentstupid.blogspot.com/2008/08/leed-vs-green-globes.html.

64. "Why Green Globes?" Green Building Initiative, https://www.thegbi.org/green-globes-certification/why-green-globes.

65. "What Is BREEAM?" BREEAM, https://www.breeam.com.

66. "BOMA BEST: Building Environmental Standards." BOMA, https://www.boma.bc.ca/green-buildings/boma-best.

67. Lance Hosey, "Six Myths of Sustainable Design," HuffPost, updated December 6, 2017, https://www.huffpost.com/entry/six-myths-of-sustainable-design_b_6823050.

68. Hosey, "Six Myths of Sustainable Design."

69. Yudelson, *Reinventing Green Building* (blog).

70. Greg Kats, Leon Alevantis, Adam Berman, Evan Mills, and Jeff Perlman, *The Costs and Financial Benefits of Green Buildings*, a report to Califor-

nia's Sustainable Building Task Force, October 2003, https://www.usgbc.org/drupal/legacy/usgbc/docs/News/News477.pdf, p. 85.

71. GSA Public Buildings Service, "Green Building Performance," GSA, August 2011, https://www.gsa.gov/cdnstatic/Green_Building_Performance2.pdf.

72. Mark Brandon, "Pros and Cons of the Green Buildings," Earthava, June 30, 2015, https://www.earthava.com/pros-and-cons-of-green-buildings.

73. Anica Landreneau, "Green Buildings Don't Have to Cost More," *Building Design + Construction*, May 2, 2017, https://www.bdcnetwork.com/blog/green-buildings-dont-have-cost-more.

74. Ilana Strauss, "This Plan for a Town Is Straight Out of a Sci-Fi Movie," From the Grapevine, January 23, 2017, https://www.fromthegrapevine.com/travel/newdealdesign-driverless-cars-moving-suburbs.

75. Arie Barendrecht, "The Future Is Now: Five Smart Building Features Transforming Today's Workplace," *Forbes*, https://www.forbes.com/sites/forbestechcouncil/2017/08/31/the-future-is-now-five-smart-building-features-transforming-todays-workplace.

76. Wikipedia, s.v. "Zero-Energy Building," last modified August 6, 2019, 18:14, https://en.wikipedia.org/wiki/Zero-energy_building.

3. BURROWING BENEATH THE EARTH

1. Wikipedia, s.v. *Blast from the Past* (Film)," last modified April 25, 2019, 21:27, https://en.wikipedia.org/wiki/Blast_from_the_Past_(film).

2. Wikipedia, s.v. *The Time Machine*," last modified August 21, 2019, 17:48, https://en.wikipedia.org/wiki/The_Time_Machine.

3. "World's Population Increasingly Urban with More Than Half Living in Urban Areas," United Nations, July 10, 2014, https://www.un.org/en/development/desa/news/population/world-urbanization-prospects-2014.html.

4. National Research Council, *Underground Engineering for Sustainable Urban Development* (Washington, DC: National Academies Press, 2013), https://www.nap.edu/catalog/14670/underground-engineering-for-sustainable-urban-development.

5. Kieron Monks, "Underground Cities: The Future of Business," CNN, November 21, 2014, http://edition.cnn.com/2014/11/18/business/underground-cities/index.html.

6. Elon Musk, "The Future We're Building—and Boring," TED Talks, April 2017, https://www.ted.com/talks/elon_musk_the_future_we_re_building_and_boring?language=en.

7. Kieran Nash, "Will We Ever . . . Live in Underground Homes?" BBC, April 21, 2015, http://www.bbc.com/future/story/20150421-will-we-ever-live-underground.

8. Nash, "Will We Ever . . . Live in Underground Homes?"

9. Nash, "Will We Ever . . . Live in Underground Homes?"

10. "The Pros and Cons of Natural and Artificial Light," Iris, https://iristech.co/the-pros-and-cons-of-natural-and-artificial-light.

11. Eric Reinholdt, "Unique Skylight Design Ideas to Bring in Sunlight and Stars," Leviton, https://home.leviton.com/blog/unique-skylight-design-ideas-to-bring-in-sunlight-and-stars (site discontinued).

12. "Groundwater Contamination." Groundwater Foundation, https://www.groundwater.org/get-informed/groundwater/contamination.html.

13. Doug McDonough, "Looking Back: 1964 World's Fair Featured Swayze Underground Home," *Plainview Herald*, October 12, 2013, https://www.myplainview.com/news/article/Looking-Back-1964-World-s-Fair-featured-Swayze-8417921.php.

14. Jen Carlson, "Is the 1960s World's Fair Underground Home Still There? An Investigation, Gothamist, March 20, 2017, https://gothamist.com/2017/03/20/underground_home_worlds_fair_revisited.php.

15. Michelle DeArmond, "Hidden Luxuries in a Fallout Shelter," *Deseret News*, May 28, 1996, https://www.deseretnews.com/article/492174/hidden-luxuries-in-a-fallout-shelter.html.

16. "How to Live Underground," Howcast, December 7, 2010, https://www.howcast.com/videos/404214-how-to-live-underground.

17. "Saving Money with Geothermal Heat Pumps," Energy Informative, https://energyinformative.org/saving-money-with-geothermal-heat-pumps.

18. Karen Graham, "Geothermal Energy: An Under-Exploited Energy Source (Part 1)," *Digital Journal*, http://www.digitaljournal.com/tech-and-science/technology/geothermal-energy-an-under-exploited-energy-source-part-1/article/484242.

19. Akshay Chavan, "Practical Advantages and Disadvantages of an Underground House," DecorDezine, April 22, 2018, https://decordezine.com/advantages-disadvantages-of-underground-house; Wikipedia, s.v. "Underground Living," last modified February 17, 2019, 23:58, https://en.wikipedia.org/wiki/Underground_living; "Pros and Cons of Building Earth Sheltered Homes," Homebuilding/Remodel Guide, https://homebuilding.thefuntimesguide.com/earth_sheltered_home; "Efficient Earth-Sheltered Homes," Energy Saver, https://www.energy.gov/energysaver/types-homes/efficient-earth-sheltered-homes.

20. Wikipedia, s.v. "Underground Living"; "Pros and Cons of Building Earth Sheltered Homes."

21. Jeff Poe Jr., "Earth-Sheltered Buildings Require Special Expertise for Roofing and Waterproofing," Terracon, January 16, 2018, https://www.terracon.com/2018/01/16/earth-sheltered-buildings-require-special-expertise-for-roofing-and-waterproofing.

22. "How to Build Underground," GreenHomeBuilding, http://www.greenhomebuilding.com/QandA/earthshelter/how.htm.

23. "Advantages and Disadvantages of Green Roofs," Green Roofers, http://www.greenroofers.co.uk/green-roofing-guides/advantages-disadvantages-green-roofs.

24. "Efficient Earth-Sheltered Homes."

25. "Raft or Mat Foundations," Understand Building Construction, http://www.understandconstruction.com/raft-foundations.html.

26. "Cost and Prices," Underground-Homes.com, http://www.underground-homes.com/cost-prices.htm.

27. "Efficient Earth-Sheltered Homes."

28. "Radon Testing," Underground-Homes.com, http://www.underground-homes.com/radon-testing.htm.

29. Valyn Daconto, "Dr. Mercola Discusses the Presence of Radon Gas in Homes," National Radon Defense (blog), June 27, 2018, https://www.nationalradondefense.com/about-us/articles/10419-dr-mercola-discusses-the-presence-of-radon-gas-in-homes.html.

30. "What Are Earth Berming and Earth Sheltering?" Solar365, http://www.solar365.com/green-homes/other/what-earth-berming-sheltering.

31. Nathan F., "How to Build an Underground, Off-Grid, Virtually Indestructible Home," https://www.offthegridnews.com/grid-threats/the-surprising-facts-about-earth-sheltered-living.

32. "History of the National Crime Syndicate," Revolvy, https://www.revolvy.com/page/History-of-the-National-Crime-Syndicate.

33. Biography.com Editors, "Enrico Fermi," Biography, last modified May 14, 2019, https://www.biography.com/scientist/enrico-fermi.

34. Wikipedia, s.v. "Lowline (Park)," last modified July 11, 2019, at 14:48, https://en.wikipedia.org/wiki/Lowline_(park).

35. David Russell Schilling, "World's First Earthscraper: 75 Story 'Inverted Pyramid' in Mexico City," *Industry Tap*, November 27, 2014, http://www.industrytap.com/worlds-first-earthscraper-75-story-inverted-pyramid-mexico-city/22909.

36. Wikipedia, s.v. "Underground City, Montreal," last modified May 28, 2019, 19:13, https://en.wikipedia.org/wiki/Underground_City,_Montreal.

37. Lisa Millar and Jack Hawke, "Helsinki's Sprawling Underground Tunnel Network Offers Shelter from Russia's Potential Threat," ABC News, July 22, 2018, https://www.abc.net.au/news/2018-07-22/helsinki-underground-tunnel-system-shelter-from-russian-threat/10022486.

38. Wikipedia, s.v. "SubTropolis," last modified April 15, 2019, 19:19, https://en.wikipedia.org/wiki/SubTropolis.

39. Gregory Scruggs, "Singapore Is Creating a Subterranean Master Plan," Next City, February 9, 2018, https://nextcity.org/daily/entry/underground-city-movement-singapore-helskini.

40. Chris Weller, "Inside the Australian Mining Town Where 80% of People Live Underground," *Business Insider*, January 27, 2016, https://www.businessinsider.com/inside-coober-pedy-australias-underground-town-2016-1.

41. Dennis Bueckert, "Moose Jaw Tunnels Reveal Dark Tales of Canada's Past," *Globe and Mail*, last modified March 23, 2018, https://www.theglobeandmail.com/life/moose-jaw-tunnels-reveal-dark-tales-of-canadas-past/article4158935.

42. Wikipedia, s.v. "North American Aerospace Defense Command," July 4, 2019, 13:26, https://en.wikipedia.org/wiki/North_American_Aerospace_Defense_Command.

43. Kaushik, "Beijing's Underground City," Amusing Planet, November 1, 2018, https://www.amusingplanet.com/2018/11/beijings-underground-city.html.

44. DHWTY, "The Incredible Subterranean City of Kish," Ancient Origins, June 20, 2014, https://www.ancient-origins.net/ancient-places-asia/incredible-subterranean-city-kish-001777.

45. Monks, "Underground Cities."

46. "Welcome to the Forestiere Underground Gardens!" Forestiere Underground Gardens, http://www.undergroundgardens.com.

47. Patrick Lynch, "Dominique Perrault's Crystalline Glass Scheme Wins Competition for Underground Multi-Modal Hub in Seoul," ArchDaily, November 1, 2017, https://www.archdaily.com/882796/dominique-perraults-crystalline-glass-scheme-wins-competition-for-underground-multi-modal-hub-in-seoul.

48. Kim Da-Sol, "Massive Underground Transit Terminal to Be Built in Gangnam," *Korea Herald*, October 23, 2017, http://www.koreaherald.com/view.php?ud=20171023000961.

4. GOING BACK TO THE SEA

1. Alison Klesman, "Record-Breaking Astronaut Peggy Whitson: 'It's an Exciting Time for Space Exploration,'" *D-brief* (blog), *Discover*, June 24, 2019, http://blogs.discovermagazine.com/d-brief/2019/06/24/record-breaking-astronaut-peggy-whitson-its-an-exciting-time-for-space-exploration; Cammy Clark, "World's Last Underwater Habitat May Fall to NOAA Budget Cut," *Seattle Times*, August 11, 2012, https://www.seattletimes.com/nation-world/worlds-last-underwater-habitat-lab-may-fall-to-noaa-budget-cut.

2. Christina Nunez, "Our Oceans Are under Attack by Climate Change, Overfishing," *National Geographic*, March 21, 2019, https://www.nationalgeographic.com/environment/habitats/ocean.

3. Clark, "World's Last Underwater-Habitat Lab May Fall to NOAA Budget Cut."

4. Wikipedia, s.v. "SEALAB," last modified August 17, 2019, 3:34, https://en.wikipedia.org/wiki/SEALAB.

5. Clark, "World's Last Underwater-Habitat Lab May Fall to NOAA Budget Cut."

6. "Jules' Undersea Lodge," https://jul.com/dive-the-lodge.

7. Rachel Nuwer, "Will We Ever . . . Live in Underwater Cities?" BBC, September 30, 2013, http://www.bbc.com/future/story/20130930-can-we-build-underwater-cities.

8. Utah State University, "Build an Ark? Biologists Discuss Conservation Prioritization," *ScienceDaily*, July 23, 2018, https://www.sciencedaily.com/releases/2018/07/180723071606.htm.

9. Ali Venosa, " Breaking Point: How Much Water Pressure Can the Human Body Take?" *Medical Daily*, August 13, 2015, https://www.medicaldaily.com/breaking-point-how-much-water-pressure-can-human-body-take-347570.

10. "Oceanography," *National Geographic*, last modified February 1, 2012, https://www.nationalgeographic.org/encyclopedia/oceanography.

11. *Sodasorb: Manual of CO_2 Absorption* (Cambridge, MA: W. R. Grace, 1993), https://www.shearwater.com/wp-content/uploads/2012/08/Sodasorb_Manual.pdf.

12. Ana Rosado, "Artificial Gills for Humans Could Become a Reality," CNN, last modified August 15, 2019, https://www.cnn.com/style/article/amphibio-underwater-breathing/index.html.

13. Wikipedia, s.v. *"Waterworld,"* last modified July 1, 2019, 18:53, https://en.wikipedia.org/wiki/Waterworld.

14. Calum Lindsay, "The Seasteading Institute's Floating Cities Are Designed for Unregulated Innovation," Dezeen, July 24, 2017, https://www.dezeen.com/2017/07/24/seasteading-institute-floating-cities-designed-for-unregulated-innovation-architecture-mini-living-initiative.

15. "Cities on the Ocean," *The Economist*, December 3, 2011, https://www.economist.com/node/21540395.

16. "Cities on the Ocean."

17. "How Do Pontoons Handle in Rough Water?" Manitou Pontoon Boats, December 21, 2017, https://www.manitoupontoonboats.com/pontoon-boats-handle-rough-water.

18. Paul Tan, "Preserving Our Ocean," *AQ31*, December 2017, http://arkdesign-architects.com/aq/AQ-31.pdf.

19. Charlie Deist, "Ocean Living: A Step Closer to Reality?" BBC, November 1, 2013, http://www.bbc.com/future/story/20131101-living-on-the-ocean.

20. Jillian Scharr, "What Ancient Roman Concrete Could Teach Modern Builders," NBC News, June 5, 2013, http://www.nbcnews.com/id/52112847/ns/technology_and_science-tech_and_gadgets/t/what-ancient-roman-concrete-could-teach-modern builders.

21. Harry Pettit, "The World's First Floating Nation Designed to 'Liberate Humanity from Politicians' Will Appear in the Pacific Ocean by 2020," *Daily Mail Online*, November 13, 2017, https://www.dailymail.co.uk/sciencetech/article-5077575/The-world-s-floating-city-set-2020-build.html.

22. Trevor Batten, "Seasteading," Tebatt.net, 2013, http://www.tebatt.net/PROJECTS/PROJECT_HOMEFARM/Project_LAND/Alternatives/SEASTEADING.html.

23. "Cities on the Ocean."

24. Brian Wang, "Startup Blue Frontiers Is Building Seastead in French Polynesia," *Nextbigfuture* (blog), November 13, 2017, https://www.nextbigfuture.com/2017/11/startup-blue-frontiers-is-building-seastead-in-french-polynesia.html.

25. Pettit, "World's First Floating Nation."

26. Mary Bellis, "Wolf Hilbertz: Sea-Cretion." The Inventors, http://theinventors.org/library/inventors/blhilbertz.htm.

27. Nuwer, "Will We Ever . . . Live in Underwater Cities?"; Emilie Chalcraft, "Water Discus Underwater Hotel by Deep Ocean Technology," Dezeen, January 29, 2013, https://www.dezeen.com/2013/01/29/worlds-largest-underwater-hotel-planned-for-dubai.

28. Sebastian Jordahn, "Ocean Spiral Is a Conceptual City Proposed beneath the Surface of the Ocean," Dezeen, November 6, 2017, https://www.dezeen.com/2017/11/06/video-ocean-spiral-shimizu-corporation-spiralling-underwater-city-movie.

29. Jordahn, "Ocean Spiral Is a Conceptual City."

30. "The WaterNest: An Eco-Friendly Floating House," AENews, April 2, 2015, http://www.alternative-energy-news.info/waternest-floating-house.

31. Sumil Chandel, "3D Printed Underwater City Would Be Made Out of Found Ocean Trash." *MyCoolBin* (blog), January 4, 2016, http://www.mycoolbin.com/2016/01/04/3d-printed-underwater-city-would-be-made-out-of-found-ocean-trash.

32. Lindsay, "The Seasteading Institute's Floating Cities."

33. Kushal Jain, "Floating Ecosystem Blue 21," Arch20, https://www.arch2o.com/floating-ecosystem-blue-21delta-sync.

34. Nuwer, "Will We Ever . . . Live in Underwater Cities?"

35. Akshata Shanbhag, "Cool Underwater Cities You Could Live In Some Day," MakeUseOf, March 16, 2015, https://www.makeuseof.com/tag/cool-underwater-cities-may-live-day.

36. Chris Kitching, "Would YOU Live in a Floating City?" *Daily Mail Online*, August 11, 2015, https://www.dailymail.co.uk/travel/travel_news/article-3192446/Would-live-floating-city-Fascinating-3-000ft-vessel-Meriens-shaped-like-giant-manta-ray-home-7-000-people-without-producing-waste.html; DNEWS, "Manta Ray-Shaped City Is a Floating University," Seeker, September 29, 2015, https://www.seeker.com/manta-ray-shaped-city-is-a-floating-university-1770284456.html; "Is This Manta Ray–Shaped Floating City the University of the Future?" *MATsolutions Test and Measurement Equipment Blog*, https://www.matsolutions.com/Blog/tabid/275/entryid/1225/Default.

37. BEC Crew, "Sure, We'd Live in This Futuristic Underwater Sphere-City," ScienceAlert, November 26, 2014, https://www.sciencealert.com/sure-we-ll-live-in-this-futuristic-underwater-sphere-city.

38. Brian Doherty, "First Seastead in International Waters Now Occupied, Thanks to Bitcoin Wealth," Reason, March 1, 2019, https://reason.com/2019/03/01/first-seastead-in-international-waters-n.

39. Adam Withnall, "Japanese Construction Firm Says This 'Ocean Spiral' Is the Underwater City of the Future," *The Independent*, November 25, 2014, https://www.independent.co.uk/news/world/asia/japanese-construction-firm-says-this-ocean-spiral-is-the-underwater-city-of-the-future-9882532.html.

40. Delana, "Futuretecture: From Sea Cities to Space Colonies," WebUrbanist, June 23, 2010, https://weburbanist.com/2010/06/23/futuretecture-from-sea-cities-to-space-colonies.

41. "Cities in the Sea," Venus Project, https://www.thevenusproject.com/resource-based-economy/environment/cities-in-the-sea.

42. Jessica Mairs, "Vincent Callebaut Proposes Underwater 'Oceanscrapers,'" Dezeen, December 24, 2015, https://www.dezeen.com/2015/12/24/aequorea-vincent-callebaut-underwater-oceanscrapers-made-from-3d-printed-rubbish-ocean-plastic.

43. "Buckminster Fuller's 40 Year Old Seastead Design," *A Place to Stand* (blog), April 4, 2009, http://a-place-to-stand.blogspot.com/2009/04/buckminster-fullers-40-year-old.html.

44. *Blue Revolution Hawaii* (blog), https://bluerevolutionhawaii.blogspot.com.

45. Bridgette Meinhold, "Underwater Ocean City for a Future Australia," Inhabitat, April 7, 2010, https://inhabitat.com/underwater-ocean-city-could-be-in-australias-future.

46. Dominique Afacan, "Apartment for Sale on Board Luxury Residential Ship the World," *Forbes*, April 23, 2017, https://www.forbes.com/sites/dominiqueafacan/2017/04/23/apartment-for-sale-on-board-luxury-residential-ship-the-world/#681f73d6edb4; "Has the Time Come for Floating Cities?" *The Guardian*, March 18, 2014, https://www.theguardian.com/cities/2014/mar/18/floating-cities-proposals-utopian-sci-fi.

47. "Has the Time Come for Floating Cities?"

48. Wikipedia, s.v. "Wave Power," last modified July 5, 2019, https://en.wikipedia.org/wiki/Wave_power.

49. Chris Woodford, "OTEC (Ocean Thermal Energy Conversion)," ExplainThatStuff! last modified December 16, 2018, https://www.explainthatstuff.com/how-otec-works.html.

50. Edward Lo, *Ocean Energy*, January 2009, ftp://ftp.ee.polyu.edu.hk/zhaoxu/EE501/OceanEnergy_intro.pdf.

51. Meghan Werft, "Is Desalination the Answer to Water Shortages?" Global Citizen, September 1, 2016, https://www.globalcitizen.org/en/content/is-desalination-the-answer-to-water-shortages.

52. Deist, "Ocean Living."

53. Doug Struck, "Treasures of the Deep: Tapping a Mineral-Rich Ocean Floor," *Trust*, August 13, 2018, https://magazine.pewtrusts.org/en/archive/summer-2018/treasures-of-the-deep-tapping-a-mineral-rich-ocean-floor; *Study to Investigate State of Knowledge of Deep-Sea Mining*, Maritime Forum, May 26, 2015, https://webgate.ec.europa.eu/maritimeforum/en/node/3732.

54. Tim Collins, "Incredible Images Reveal a Futuristic Vision," *Daily Mail Online*, December 18, 2017, https://www.dailymail.co.uk/sciencetech/article-5190277/Concept-images-reveal-worlds-floating-nation.html; Josh Gabbatiss, "World's First Floating City to Be Built off the Coast of French Polynesia by 2020," *The Independent*, November 14, 2017, https://www.independent.co.uk/news/science/floating-city-french-polynesia-2020-coast-islands-south-pacific-

ocean-peter-thiel-seasteading-a8053836.html; Donna Maria Padua, "The World Will See the First Floating Nation in the Pacific Ocean," Elite Readers, https://www.elitereaders.com/world-first-floating-nation-pacific-ocean.

5. SETTLEMENTS ABOVE THE SKY

1. Wikipedia, s.v. "Outer Space Treaty," last modified August 6, 2019, 10:23, https://en.wikipedia.org/wiki/Outer_Space_Treaty.

2. "Outer Space Treaty," [Treaty on Principles Governing the Activities of States in the Exploration and Use of Outer Space, Including the Moon and Other Celestial Bodies,] January 27, 1967, US Department of State, https://2009-2017.state.gov/t/isn/5181.htm#signatory.

3. Erin Durkin, "Space Force: All You Need to Know about Trump's Bold New Interstellar Plan," *The Guardian*, August 10, 2018, https://www.theguardian.com/us-news/2018/aug/10/space-force-everything-you-need-to-know.

4. Wikipedia, s.v. "United States Space Force," last modified August 21, 2019, 11:36, https://en.wikipedia.org/wiki/United_States_Space_Force.

5. Victor Tangermann, "Trump Signs Order to Create a U.S. Space Command," Futurism, December 18, 2018, https://futurism.com/trump-executive-order-space-command-space-force.

6. "Trump's 'Space Command' Comes One Step Closer to Existence," *New York Post*, December 12, 2017, https://nypost.com/2018/12/17/trumps-space-command-comes-one-step-closer-to-existence.

7. Tangermann, "Trump Signs Order to Create a U.S. Space Command."

8. "Outer Space Treaty."

9. Rand Simberg, "Property Rights in Space," *New Atlantis*, Fall 2012, https://www.thenewatlantis.com/publications/property-rights-in-space.

10. Bilzin Sumberg, "It's Up in the Air: Air Rights in Modern Development," Lexology, March 26, 2015, https://www.lexology.com/library/detail.aspx?g=a89b4a0a-1cb5-461d-96cc-13c76e7d75be.

11. Wikipedia, s.v. "Extraterrestrial Real Estate," last modified July 9, 2019, 12:33, https://en.wikipedia.org/wiki/Extraterrestrial_real_estate.

12. *Mars, National Geographic*, https://www.nationalgeographic.com/tv/mars.

13. Tim Delaney and Tim Madigan, *Beyond Sustainability* (Jefferson, NC: McFarland & Co., 2014).

14. Wikipedia, s.v. "Where No Man Has Gone Before," last modified July 8, 2019, 14:05, https://en.wikipedia.org/wiki/Where_no_man_has_gone_before.

15. Chelsea Gohd, "Stephen Hawking: Humans Must Leave Earth within 600 Years," Futurism, November 7, 2017, https://futurism.com/stephen-hawking-humans-must-leave-earth-within-600-years.

16. Kristin Houser, "Bill Nye on Terraforming Mars: 'Are You Guys High?'" Futurism, November 19, 2018, https://futurism.com/bill-nye-mars-terraforming-high.

17. Jolene Creighton, "Neil deGrasse Tyson Says Humans Will Never Colonize Mars," Futurism, February 20, 2018, https://futurism.com/neil-degrasse-tyson-humans-colonize-mars.

18. Ben Austen, "After Earth: Why, Where, How, and When We Might Leave Our Home Planet," Popular Science, March 16, 2011, https://www.popsci.com/science/article/2011-02/after-earth-why-where-how-and-when-we-might-leave-our-home-planet.

19. Wikipedia, s.v. "Space Colonization," last modified July 17, 2019, 12:17, https://en.wikipedia.org/wiki/Space_colonization.

20. Victoria Jaggard, "Toxic Mars Dust Could Hamper Planned Human Missions," New Scientist, May 8, 2013, https://www.newscientist.com/article/dn23505-toxic-mars-dust-could-hamper-planned-human-missions.

21. Adrian Higgins, "Humans Have Been Using Their Waste as Fertilizer for Centuries," Washington Post, August 23, 2017, https://www.washingtonpost.com/lifestyle/home/would-you-use-human-waste-in-your-garden/2017/08/22/43556b90-82b4-11e7-b359-15a3617c767b_story.html.

22. Tia Ghose, "'The Martian': What Would It Take to Grow Food on Mars?" Live Science, October 9, 2015, https://www.livescience.com/52444-growing-food-on-mars.html.

23. Mike Wall, "Mars Cave-Exploration Mission Entices Scientists," Space.com, November 20, 2012, https://www.space.com/18546-mars-caves-sample-return-mission.html.

24. Andrew Chaikin, "Is SpaceX Changing the Rocket Equation?" Air & Space Magazine, January 2012, https://www.airspacemag.com/space/is-spacex-changing-the-rocket-equation-132285884.

25. D. V. Smitherman Jr., "Space Elevators," National Space Society, August 2000, https://space.nss.org/media/2000-Space-Elevator-NASA-CP210429.pdf.

26. Tauri Group, NASA Socio-Economic Impacts, NASA, https://www.nasa.gov/sites/default/files/files/SEINSI.pdf.

27. Matt Williams, "How Do We Terraform Mars?" Universe Today. March 15, 2016, https://www.universetoday.com/113346/how-do-we-terraform-mars.

28. NASA, "Subsurface Explorers," Mars Exploration, https://mars.nasa.gov/programmissions/missions/missiontypes/subsurface.

29. Matt Wall, "Incredible Technology: NASA's Wild Airship Idea for Cloud Cities on Venus," Space.com, April 20, 2015, https://www.space.com/29140-venus-airship-cloud-cities-incredible-technology.html.

30. Jessa Gamble, "How Do You Build a City in Space?" The Guardian, May 16, 2014, https://www.theguardian.com/cities/2014/may/16/how-build-city-in-space-nasa-elon-musk-spacex.

31. Gamble, "How Do You Build a City in Space?"

32. "How Much Radiation Will the Settlers Be Exposed To?" Mars One, https://www.mars-one.com/faq/health-and-ethics/how-much-radiation-will-the-settlers-be-exposed-to.

33. Danny Lewis, "Future Moon Bases Might Be Built in Underground Lava Tubes," The Smithsonian, March 29, 2016, https://www.smithsonianmag.

com/smart-news/future-moon-bases-might-be-built-underground-lava-tubes-180958590.

34. Wikipedia, s.v. "Space Manufacturing," last modified July 15, 05:05, https://en.wikipedia.org/wiki/Space_manufacturing.

35. Swapna Krishna, "Russia and the US Will Work Together to Build a Moon Base," Engadget, September 27, 2017, https://www.engadget.com/2017/09/27/russia-us-cooperate-on-lunar-base.

36. Leah Crane, "Terraforming Mars Might Be Impossible Due to a Lack of Carbon Dioxide," *New Scientist*, July 30, 2018, https://www.newscientist.com/article/2175414-terraforming-mars-might-be-impossible-due-to-a-lack-of-carbon-dioxide.

37. Jacob Banas, "NASA Clears 'Dream Chaser' Space Cargo Plane for Full-Scale Production," Futurism, December 30, 2018, https://futurism.com/the-byte/nasa-dream-chaser-space-cargo-plane.

38. Sofie Curtis, "NASA Wants to Put a Magnetic Shield around Mars so that Humans Can Live on the Red Planet," *The Mirror*, March 6, 2017, https://www.mirror.co.uk/science/nasa-wants-put-magnetic-shield-9973347.

39. Victor Tangermann, "Off World: A Timeline for Humanity's Colonization of Space," Futurism, October 30, 2017, https://futurism.com/a-timeline-for-humanitys-colonization-of-space.

40. Tom McCarthy, "NASA Lays Out Vision for Manned Mission to Mars—As It Happened," *The Guardian*, May 6, 2013, https://www.theguardian.com/science/2013/may/06/nasa-curiosity-mars-press-conference-live.

41. Mariella Moon, "Jeff Bezos Is Planning a Delivery Service for the Moon," Engadget, March 3, 2017, https://www.engadget.com/2017/03/03/jeff-bezos-delivery-service-moon-blue-origin.

42. Banas, "NASA Clears 'Dream Chaser' Space Cargo Plane for Full-Scale Production."

43. "UAE Launches Space Program to Boost Colonization of Mars by 2021," RT, April 13, 2017, https://www.rt.com/news/384638-uae-space-program-mars-colonization.

44. Steven Jiang "China Launches Satellite to Explore Moon's Dark Side," WSLS, May 21, 2018, https://www.wsls.com/tech/china-launches-satellite-to-explore-moons-dark-side.

45. "Vision for Space Exploration_Negative_HHM_SDI_2011," National Debate Coaches Association, http://open-evidence.s3-website-us-cast-1.amazonaws.com/2011.html.

46. "Greenhouse Effects . . . Also on Other Planets," ESA, February 14, 2003, https://www.esa.int/Our_Activities/Space_Science/Venus_Express/Greenhouse_effects_also_on_other_planets.

47. Daniel Oberhaus and Alex Pasternack, "Why We Should Build Cloud Cities on Venus," February 5, 2015, Vice, https://www.vice.com/en_us/article/539jj5/why-we-should-build-cloud-cities-on-venus.

48. "Venus," NASA Science, last modified April 5, 2019, https://solarsystem.nasa.gov/planets/venus/in-depth.

49. Brandon Weigel, "A Colony on Venus," Medium, April 9, 2018, https://medium.com/our-space/a-colony-on-venus-994182f3ea41.

50. A. Crowl, "Terraforming Venus: A Comparison of Methods," Crowl-space, http://crowlspace.com/?p=1959.

51. "Why Explore Venus?" Venus Labs, http://www.venuslabs.org/?p=explore; Adam Becker, "The Amazing Cloud Cities We Could Build on Venus," BBC News, October 20, 2016, http://www.bbc.com/future/story/20161019-the-amazing-cloud-cities-we-could-build-on-venus.

52. "The Asteroid Belt," NASA, https://starchild.gsfc.nasa.gov/docs/StarChild/solar_system_level2/asteroids.html.

53. Harry Pettit, "NASA Headed Towards Giant Golden Asteroid That Could Make Everyone on Earth a Billionaire," Fox News, June 27, 2019, https://www.foxnews.com/science/nasa-headed-towards-giant-golden-asteroid-that-could-make-everyone-on-earth-a-billionaire.

54. Al Globus, "Space Settlement Basics," National Space Society, https://space.nss.org/settlement/nasa/Basics/wwwwh.html.

55. Nola Taylor Redd, "Jupiter's Icy Moon Europa: Best Bet for Alien Life?" Space.com, August 22, 2014, https://www.space.com/26905-jupiter-moon-europa-alien-life.html.

56. Cory Scarola, "Is It Possible to Float a Skyscraper from an Asteroid?" Inverse, March 31, 2017, https://www.inverse.com/article/29789-floating-skyscraper-asteroid-impossible-analemma-tower.

57. Katharine Gammon, "Terrestrial Planets: Definition & Facts about the Inner Planets," Space.com, February 8, 2019, https://www.space.com/17028-terrestrial-planets.html.

58. C. A. Evans et al., "International Space Station Science Research Accomplishments during the Assembly Years: An Analysis of Results from 2000–2008," NASA, June 2009, https://www.nasa.gov/pdf/389388main_ISS%20Science%20Report_20090030907.pdf.

59. Jerry Coffey, "Temperature of Mars," Universe Today, June 7, 2008, https://www.universetoday.com/14911/temperature-of-mars.

60. Ivan Couronne, "After the Moon, People on Mars by 2033 . . . or 2060," Phys.org, May 18, 2019, https://phys.org/news/2019-05-moon-people-mars.html.

61. Dan Nosowitz, "Scientists Grow Vegetables on 'Martian' Soil; Believe We Won't Die from Eating Them," Modern Farmer, June 30, 2016, https://modernfarmer.com/2016/06/safe-to-eat-food-on-mars.

62. Sonica Krishan, "Why NASA Has Recommends Spirulina as Excellent Space Food?" Health Blog, Sanat Products Ltd., https://www.sanat.co.in/health-blog/123/why-nasa-has-recommends-spirulina-as-excellent-space-food.

63. Shaunacy Ferro, "How Do Astronauts Get Drinking Water on the ISS?" Mental Floss, September 7, 2015, http://mentalfloss.com/article/67854/how-do-astronauts-get-drinking-water-iss.

64. Charles W. Dunnill, "Turning Water into Oxygen in Zero Gravity Could Mean Easier Trips to Mars," Popular Science, July 13, 2018, https://www.popsci.com/making-oxygen-from-water-in-space.

65. "Breathing Easy on the Space Station," NASA Science, https://science.nasa.gov/science-news/science-at-nasa/2000/ast13nov_1.

66. Associated Press, "Scientists Say Mars Soil Contains Chemical Found in Rocket Fuel, Fireworks," *Mercury News*, August 4, 2008, https://www.mercurynews.com/2008/08/04/scientists-say-mars-soil-contains-chemical-found-in-rocket-fuel-fireworks.

67. Neel V. Patel, "Weather Got You Down? The Entire Planet of Mars Is Buried in a Dust Storm Right Now," *Popular Science*, June 22, 2018, https://www.popsci.com/mars-global-dust-storm.

68. "The Luna Ring Concept," SSERVI, https://sservi.nasa.gov/articles/the-luna-ring-concept.

69. "What Is Solar Sailing?" Planetary Society, http://www.planetary.org/explore/projects/lightsail-solar-sailing/what-is-solar-sailing.html.

70. Delana, "Futuretecture: From Sea Cities to Space Colonies," Web Urbanist, June 23, 2010, https://weburbanist.com/2010/06/23/futuretecture-from-sea-cities-to-space-colonies.

71. Globus, "Space Settlement Basics."

72. Sarah Knapton, "Elon Musk: We'll Create a City on Mars with a Million Inhabitants," *The Telegraph*, June 21, 2017, https://www.telegraph.co.uk/science/2017/06/21/elon-musk-create-city-mars-million-inhabitants.

73. Sarah Knapton, "Elon Musk Reveals Plans for 'Fun' Mars City: 1m People and Pizza Joints within 50 Years," *Sidney Morning Herald*, June 22, 2017, https://www.smh.com.au/technology/elon-musk-reveals-plans-for-fun-mars-city-1m-people-and-pizza-joints-20170622-gww1n8.html.

74. Sarah Knapton, "NASA Planning 'Earth Independent' Mars Colony by 2030s," *The Telegraph*, October 9, 2015, https://www.telegraph.co.uk/science/2016/03/14/nasa-planning-earth-independent-mars-colony-by-2030s.

75. Sarah Fecht, "Bricks Made from Fake Martian Soil Are Surprisingly Strong," *Popular Science*, April 27, 2017, https://www.popsci.com/mars-soil-bricks.

76. Colin Lecher, "Now on Kickstarter: The First Steps toward a Lunar Space Elevator," *Popular Science*, August 27, 2012, https://www.popsci.com/technology/article/2012-08/group-launches-kickstarter-take-first-step-toward-space-elevator.

77. Scott Snowden, "Japanese Space-Elevator Experiment Launching to Space Station Next Week (Really!)," Space.com, September 7, 2018, https://www.space.com/41278-japan-space-elevator-cubesats-experiment.html.

78. Scott Snowden, "A Colossal Elevator to Space Could Be Going Up Sooner Than You Ever Imagined," NBC News, October 2, 2018, https://www.nbcnews.com/mach/science/colossal-elevator-space-could-be-going-sooner-you-ever-imagined-ncna915421; Wikipedia, s.v. "Space Elevator," last modified June 25, 2019, 21:42, https://en.wikipedia.org/wiki/Space_elevator; Graham Templeton, "60,000 Miles Up: Space Elevator Could Be Built by 2035, Says New Study," ExtremeTech, March 6, 2014, https://www.extremetech.com/extreme/176625-60000-miles-up-geostationary-space-elevator-could-be-built-by-2035-says-new-study; Elizabeth Rayne, "A Space Elevator Is Actually

Happening," Syfy Wire, October 11, 2018, https://www.syfy.com/syfywire/a-space-elevator-is-actually-happening.

79. Wikipedia, s.v. "DirecTV," last modified June 24, 2019, 17:04, https://en.wikipedia.org/wiki/DirecTV.

80. Mary Roach, *Packing for Mars* (New York: W. W. Norton & Company, 2010).

81. Ryan Bradley, "Can Billionaire Robert Bigelow Create a Life for Humans in Space?" *Popular Science*, April 8, 2016, https://www.popsci.com/can-billionaire-robert-bigelow-create-a-life-for-humans-in-space.

82. Roach, *Packing for Mars*.

83. Loren Grush, "Virgin Galactic's Spaceplane Finally Makes It to Space for the First Time," The Verge, December 13, 2018, https://www.theverge.com/2018/12/13/18138279/virgin-galactic-vss-unity-spaceshiptwo-space-tourism.

84. Kristin Houser, "Virgin Galactic Will Send People to Space by Christmas. Maybe," Futurism, November 30, 2018, https://futurism.com/virgin-galactic-branson-christmas.

85. Ian Sample. "Virgin Galactic Space Shot Is Go 'Within Weeks, Not Months,'" *The Guardian*, October 9, 2018, https://www.theguardian.com/science/2018/oct/09/virgin-galactic-space-flight-vss-unity-go-weeks-not-months.

86. Jeff Foust, "Space Resources Company Co-Founder Sets Sights on Next Wave of Space Startups," *Space News*, April 25, 2018, https://spacenews.com/space-resources-company-co-founder-sets-sights-on-next-wave-of-space-startups.

87. Andrew Jones, "China's Mission to the Far Side of the Moon Will Launch in December," *Guest Blog*, Planetary Society, August 16, 2018, http://www.planetary.org/blogs/guest-blogs/change4-launch-announced.html.

88. Daniel Oberhaus, "The Never-Ending Quest to Build a Hotel in Space," The Outline, February 13, 2017, https://theoutline.com/post/1073/hilton-never-ending-quest-to-build-a-hotel-in-space.

89. Richard Hollingham, "Building a New Society in Space," BBC Future, March 18, 2013, http://www.bbc.com/future/story/20130318-building-a-new-society-in-space.

90. Dan Robitzski, "Antibiotic-Resistant Bacteria Found on International Space Station," Futurism, November 27, 2018, https://futurism.com/antibiotic-resistant-bacteria-international-space-station.

91. "George Santayana Quote about History," Wisdom Quotes, http://wisdomquotes.com/learn-from-history-george-santayana.

6. THE MOST IMPORTANT COLOR
IN SUPERCITIES

1. Jill Suttie, "How Nature Can Make You Kinder, Happier, and More Creative," *Greater Good*, March 2, 2016, https://greatergood.berkeley.edu/article/item/how_nature_makes_you_kinder_happier_more_creative.

2. Oliver Balch, "Garden Cities: Can Green Spaces Bring Health and Happiness," *The Guardian*, April 20, 2015, https://www.theguardian.com/sustainable-business/2015/apr/20/garden-cities-can-green-spaces-bring-health-and-happiness; M. Braubach et al., "Effects of Urban Green Space on Environmental Health, Equity. and Resilience," in *Nature-Based Solutions to Climate Change Adaptation in Urban Areas: Theory and Practice of Urban Sustainability Transitions*, ed. N. Kabisch, H. Korn, J. Stadler, and A. Bonn (Cham, Switzerland: Springer, 2017), 187–205; "Active Living," Green Cities: Good Health, College of the Environment, University of Washington, 2010, https://depts.washington.edu/hhwb/Thm_ActiveLiving.html; A. C. K. Lee, H. C. Jordan, J. Horsley, "Value of Urban Green Spaces in Promoting Healthy Living and Wellbeing," *Risk Management and Healthcare Policy* 8, 2015: 131–37.

3. Institute of Health Equity, *Improving Access to Green Spaces*, https://assets.publishing.service.gov.uk/government/uploads/system/uploads/attachment_data/file/355792/Briefing8_Green_spaces_health_inequalities.pdf.

4. Bianca Barragan, "11-Mile Los Angeles River Restoration Could Move Forward This Fall," Curbed Los Angeles, August 22, 2017, https://la.curbed.com/2017/8/22/16181314/los-angeles-river-restoration-move-forward.

5. Alyssa Ackerman, "Get Outside: Portland's Most Beautiful Parks and Green Spaces," Fitt Portland, April 18, 2019, https://fitt.co/portland/articles/parks-portland.

6. Simon Weedy, "Students Creating 'Islands of Cool' for Paris Schoolyards in Summer," Child in the City, August 27, 2018, https://www.childinthecity.org/2018/08/27/students-creating-islands-of-cool-for-paris-schoolyards-in-summer.

7. Laura Testino, "The Allure of Vertical Forests," *New York Times*, December 13, 2018, https://www.nytimes.com/2018/12/13/us/stefano-boeri-vertical-forests-cities-conference.html.

8. Testino, "The Allure of Vertical Forests."

9. Abigail Abrams, "What Green Spaces and Nature Can Do to Your Mood," *Time*, August 7, 2017, https://time.com/4881665/green-spaces-nature-happiness.

10. Testino, "The Allure of Vertical Forests."

11. J. Barton and M. Rogerson, "The Importance of Greenspace for Mental Health," *BJPsych International* 14, no. 4 (November 2017): 79–81, https://www.ncbi.nlm.nih.gov/pmc/articles/PMC5663018.

12. Abrams, "What Green Spaces and Nature Can Do to Your Mood."

13. "Urban Planning and the Importance of Green Space in Cities to Human and Environmental Health," Healthy Parks, Healthy People Central, https://www.parksforcalifornia.org.

14. Gretchen Reynolds, "How Walking in Nature Changes the Brain," *Well* (blog), *New York Times*, July 22, 2015, https://well.blogs.nytimes.com/2015/07/22/how-nature-changes-the-brain.

15. Kent Allen Halliburton, "How Green Space Benefits Our Wellbeing," Save the Earth, January 16, 2017, https://savetheearth.coop/blog/green-space-benefits.

16. Reynolds, "How Walking in Nature Changes the Brain."

17. Megan Lewis, "How Cities Use Parks for Economic Development," American Planning Association, April 1, 2003, https://www.planning.org/publications/document/9148668.

18. "Why City Parks Matter," City Parks Alliance, https://cityparksalliance.org/about-us/why-city-parks-matter.

19. "Basics of Heat Loss, Heat Gain," *ACHR News*, https://www.achrnews.com/articles/83175-basics-of-heat-loss-heat-gain.

20. Steph, "Green City Rehab: 12 Eco Urban Makeover Concepts," Web Ecoist, https://www.momtastic.com/webecoist/2010/12/13/green-city-rehab-12-eco-urban-makeover-concepts.

21. Steph, "Green City Rehab."

22. "Mental Health & Function," Green Cities: Good Health, College of the Environment, University of Washington, 2010, https://depts.washington.edu/hhwb/Thm_Mental.html.

23. "Extra Credit Built Environment Analysis," GSU, April 29, 2016, http://sites.gsu.edu/lirving1/2016/04/29/extra-credit-built-environment-analysis.

24. Ming Kuo, "Six Ways Nature Helps Children Learn," *Greater Good*, June 7, 2019, https://greatergood.berkeley.edu/article/ITEM/six_ways_nature_helps_children_learn.

25. Bright Horizons Education Team, "Benefits of Nature for Kids," Bright Horizons, https://www.brighthorizons.com/family-resources/children-and-nature.

26. "Healing," Green Cities: Good Health, College of the Environment, University of Washington, 2010, https://depts.washington.edu/hhwb/Thm_Healing.html.

27. M. Braubach et al., "Effects of Urban Green Space on Environmental Health, Equity. and Resilience."

28. "Mental Health & Function."

29. Farida Perveen, "Effects of Horticulture Therapy for Elderly with Dementia in an Institutional Setting: A Literature Review," (degree thesis, Arcada, 2013), http://docplayer.net/10699927-Effects-of-horticulture-therapy-for-elderly-with-dementia-in-an-institutional-setting.html.

30. "Roger S. Ulrich, Ph.D., EDAC," Center for Health Design, https://www.healthdesign.org/about-us/meet-team/roger-s-ulrich-phd-edac.

31. Alexandra Sifferlin, "How Nature Can Help Prisoners," *Time*, August 5, 2016, https://time.com/4440974/nature-forest-prison-inmates.

32. "Mental Health by the Numbers," NAMI: National Alliance on Mental Health, https://www.nami.org/learn-more/mental-health-by-the-numbers.

33. "The Importance of Nature in Older Populations," Nature Sacred, August 2, 2016, https://naturesacred.org/nature-seniors.

34. Eryn Pleson et al., "Understanding Older Adults' Usage of Community Green Spaces in Taipei, Taiwan," *International Journal of Environmental Research and Public Health* 11, no. 2 (February 2014): 1444–64, https://www.ncbi.nlm.nih.gov/pmc/articles/PMC3945547.

35. Emma Wood et al., "Not All Green Space Is Created Equal: Biodiversity Predicts Psychological Restorative Benefits From Urban Green Space," *Frontiers in Psychology*, November 27, 2018, https://www.frontiersin.org/articles/10.3389/fpsyg.2018.02320/full.

36. K. Lachowycz and A. P. Jones, "Greenspace and Obesity: A Systematic Review of the Evidence," ResearchGate, originally published February 2011, https://www.researchgate.net/publication/50195000_Greenspace_and_obesity_A_systematic_review_of_the_evidence.

37. Dina Castro and Ann Harman, "Growing Healthy Kids: A Community Garden–Based Obesity Program," ResearchGate, originally published March 2018, https://www.researchgate.net/publication/235647380_Growing_Healthy_Kids_A_Community_Garden-Based_Obesity_Prevention_Program.

38. M. Braubach et al., "Effects of Urban Green Space on Environmental Health, Equity, and Resilience."

39. Sally Robertson, "Stroke Prognosis," News-Medical.Net, last updated August 23, 2018, https://www.news-medical.net/health/Stroke-Prognosis.aspx.

40. Florence Williams, "Call to the Wild: This Is Your Brain on Nature," *National Geographic*, https://www.nationalgeographic.com/magazine/2016/01/call-to-wild.

41. Meg Selig, "What Did You Say?! How Noise Pollution Is Harming You," *Psychology Today*, September 25, 2013, https://www.psychologytoday.com/us/blog/changepower/201309/what-did-you-say-how-noise-pollution-is-harming-you.

42. May Wong, "Stanford Study Finds Walking Improves Creativity," *Stanford News*, April 24, 2014, https://news.stanford.edu/2014/04/24/walking-vs-sitting-042414.

43. "Mental Health & Function."

44. "Ecotherapy/Nature Therapy," GoodTherapy, https://www.goodtherapy.org/learn-about-therapy/types/econature-therapy.

45. Lawrence Robinson, Jeanne Segal, and Melinda Smith, "The Mental Health Benefits of Exercise," HelpGuide, last modified June 2019, https://www.helpguide.org/articles/healthy-living/the-mental-health-benefits-of-exercise.htm.

46. J. K. Summers and D. N. Vivian, "Ecotherapy—A Forgotten Ecosystem Service: A Review," *Frontiers in Psychology*, August 3, 2018, https://www.ncbi.nlm.nih.gov/pmc/articles/PMC6085576.

47. "Mental Health & Function."

48. Rebecca A. Clay, "Green Is Good for You," *Monitor on Psychology*, American Psychological Association, April 2001, https://www.apa.org/monitor/apr01/greengood.aspx.

49. J. A. Blumenthal, P. J. Smith, and B. M. Hoffman, "Is Exercise a Viable Treatment for Depression?" HHS Public Access, July 1, 2013, https://www.ncbi.nlm.nih.gov/pmc/articles/PMC3674785.

50. Barton and Rogerson, "The Importance of Greenspace for Mental Health."

51. Summers and Vivian, "Ecotherapy—A Forgotten Ecosystem Service."

52. William Browning, Catherine Ryan, and Joseph Clancy, *14 Patterns of Biophilic Design* (New York: Terrapin Bright Green, LLC, 2014), https://www.lbhf.gov.uk/sites/default/files/section_attachments/14_patterns_of_biophilic_design_-_improving_health_well-being_in_the_built_environment.pdf.

53. Braubach, "Effects of Urban Green Space on Environmental Health, Equity, and Resilience."

54. Patrick Sisson, "The Landscape Architect Who Helped Invent Modern City Parks," Curbed, November 23, 2016, https://www.curbed.com/2016/11/22/13712802/landscape-architecture-lawrence-halprin.

55. Wikipedia, s.v. "High Line," last modified July 15, 2019, 19:16, https://en.wikipedia.org/wiki/High_Line.

56. Sisson, "The Landscape Architect Who Helped Invent Modern City Parks."

57. "Privately-Owned Public Open Space and Public Art," San Francisco Planning, https://sfplanning.org/privately-owned-public-open-space-and-public-art.

58. "5 Key Factors in Urban Planning." EagleView, last modified February 2019, https://www.eagleview.com/2016/08/5-key-factors-in-urban-planning.

59. Christopher Walker, "Partnerships for Parks," Urban Institute, April 1, 1999, https://www.urban.org/research/publication/partnerships-parks.

60. Braubach, "Effects of Urban Green Space on Environmental Health, Equity, and Resilience."

7. BLEEDING-EDGE BUILDING MATERIALS

1. "3D Printing in Construction," Designing Buildings Wiki, July 26, 2019, https://www.designingbuildings.co.uk/wiki/3D_printing_in_construction.

2. Rory Stott, "Chinese Company Showcases Ten 3D-Printed Houses," ArchDaily, September 2, 2014, https://www.archdaily.com/543518/chinese-company-showcases-ten-3d-printed-houses.

3. Romain de Laubier et al., "Will 3D Printing Remodel the Construction Industry?" BCG, January 23, 2018, https://www.bcg.com/en-us/publications/2018/will-3d-printing-remodel-construction-industry.aspx.

4. John Norton, *Building with Earth: A Handbook* (Rugby, UK: Intermediate Technology Publications, 1986), http://library.uniteddiversity.coop/Ecological_Building/Building_With_Earth-A_Handbook.pdf.

5. Wikipedia, s.v. "Contour Crafting," last modified July 9, 2019, 09:12, https://en.wikipedia.org/wiki/Contour_crafting.

6. Kaya Yurieff, "This Robot Can 3D Print a Building in 14 Hours," CNN Business, May 2, 2017, https://money.cnn.com/2017/05/02/technology/3d-printed-building-mit/index.html.

7. Chris Mills, "This $10,000 3D Printed House Can Be Built in 24 Hours and Is Bigger Than a Studio Apartment," BGR, March 12, 2018, https://bgr.com/2018/03/12/3-d-printed-house-icon-sxsw.

8. "Wire + Arc Additive Manufacturing," WAAM, https://waammat.com/about/waam.

9. David Tal, "Housing Prices Crash as 3D Printing and Maglevs Revolutionize Construction," Quantumrun, https://www.quantumrun.com/prediction/housing-prices-crash-3d-printing-and-maglevs-revolutionize-construction-future-cities-p3.

10. "Chinese Construction Firm Erects 57-Storey Skyscraper in 19 Days," *The Guardian*, April 30, 2015, https://www.theguardian.com/world/2015/apr/30/chinese-construction-firm-erects-57-storey-skyscraper-in-19-days.

11. Nikita Cheniuntai, "What Is Construction 3D Printing: Perspectives and Challenges," Medium, January 29, 2018, https://medium.com/@Nik_chen/what-is-construction-3d-printing-perspectives-and-challanges-5b57170c2a29.

12. De Laubier et al., "Will 3D Printing Remodel the Construction Industry?"

13. Whirlwind Team, "Impacts of 3D Printing on the Construction Industry," *Whirlwind Steel* (blog), March 30, 2016, https://www.whirlwindsteel.com/blog/impacts-of-3d-printing-on-the-construction-industry.

14. "Remodeling Construction Industry with 3d Enhancement," *The Masterbuilder*, July 9, 2018, https://www.masterbuilder.co.in/?p=83605.

15. Folkert Haag, "How 3D Printing Is Disrupting the Building Materials Industry," *Forbes*, August 20, 2018, https://www.forbes.com/sites/sap/2018/08/20/how-3d-printing-is-disrupting-the-materials-building-industry.

16. De Laubier et al., "Will 3D Printing Remodel the Construction Industry?"

17. Anwen Haynes, "The Pros and Cons of 3D Printing," Mode Solutions, June 19, 2017, https://www.modeprintsolutions.co.uk/2017/06/the-pros-and-cons-of-3d-printing; Crystal Ayres, "13 Pros and Cons of 3D Printing," *Green Garage* (blog), July 19, 2016, https://greengarageblog.org/13-pros-and-cons-of-3d-printing.

18. Wikipedia, s.v. "Aerographite," last modified December 23, 2016, 21:11, https://en.wikipedia.org/wiki/Aerographite.

19. Talha Dar, "Airloy Is the New Super Material of the Future That Is 100 Times Lighter Than Water," Wonderful Engineering, 2015, https://wonderfulengineering.com/airloy-is-the-new-super-material-of-the-future-that-is-100-times-lighter-than-water.

20. Wikipedia, s.v. "Metal Foam," last modified July 20, 2019, 22:17, https://en.wikipedia.org/wiki/Metal_foam.

21. Anthony Watts, "Useful: Making Concrete from Coal Ash," Watts Up with That? July 12, 2018, https://wattsupwiththat.com/2018/07/12/useful-making-concrete-from-coal-ash.

22. Connor Walker, "Bamboo: A Viable Alternative to Steel Reinforcement?" ArchDaily, June 8, 2014, https://www.archdaily.com/513736/bamboo-a-viable-alternative-to-steel-reinforcement.

23. US Biochar Initiative, https://biochar-us.org.

24. Rice University, "Researchers Enhance Boron Nitride Nanotubes for Next-Gen Composites," Phys.org, May 21, 2018, https://phys.org/news/2018-05-boron-nitride-nanotubes-next-gen-composites.html.

25. University at Buffalo, "Carbon Nanotubes Are Superior to Metals for Electronics, According to Engineers," ScienceDaily, March 24, 2009, https://www.sciencedaily.com/releases/2009/03/090320134041.htm.

26. Ganesh Ramanathan, "Color Changing Technology for Cars Made Easier with Spectrophotometric Paint Analysis," *Blog*, HunterLab, January 18, 2017, https://www.hunterlab.com/blog/color-chemical-industry/color-changing-technology-for-cars-with-spectrophotometric-paint-analysis.

27. Andrew Michler, "New CO2 Sand Bricks Are 2.5 Times Stronger Than Concrete," Inhabitat, August 10, 2011, https://inhabitat.com/new-co2-sand-bricks-are-2-5-times-stronger-than-concrete.

28. Marlene Cimons, "Transforming Waste in Order to Transform People's Lives," NSF, July 22, 2014, https://www.nsf.gov/discoveries/disc_summ.jsp?cntn_id=132101&org=NSF.

29. Jiang He and Akira Hoyano, "Experimental Study of Cooling Effects of a Passive Evaporative Cooling Wall Constructed of Porous Ceramics with High Water Soaking-Up Ability," *Building and Environment* 45, no. 2 (February 2010): 461–72.

30. James Vincent, "'Five-Dimensional' Glass Discs Can Store Data for Up to 13.8 Billion Years," The Verge, February 16, 2016, https://www.theverge.com/2016/2/16/11018018/5d-data-storage-glass.

31. Nathan Nakahara and Celine Chen, "Sustainable Innovation in New Materials," Plug and Play, https://www.plugandplaytechcenter.com/resources/sustainable-innovation-new-materials.

32. A.Jacks delightus peter, D. Balaji, and D. Gowrishankar, "Waste Heat Energy Harvesting Using Thermo Electric Generator," *IOSR Journal of Engineering* 3, no. 7 (July 2013): 1–4, https://www.iosrjen.org/Papers/vol3_issue7%20(part-2)/A03720104.pdf.

33. Graham Templeton, "Ferrock: A Carbon Dioxide Sponge That's Harder Than Concrete," Geek, November 14, 2014, https://www.geek.com/news/ferrock-a-carbon-dioxide-sponge-thats-harder-than-concrete-1609410.

34. "Black Graphene Radiation Protecting Paint," Building Centre, https://www.buildingcentre.co.uk/supermaterial/black-graphene-radiation-protecting-paint.

35. Jessica Hulinger, "Metal as Light as Styrofoam and as Strong as Titanium—and 5 Other Amazing New Materials," *The Week*, October 20, 2015,

https://theweek.com/articles/582787/metal-light-styrofoam-strong-titanium--5-other-amazing-new-materials.

36. "Bulletproof Graphene Makes Ultra-Strong Body Armour," *New Scientist*, November 27, 2014, https://www.newscientist.com/article/dn26626-bulletproof-graphene-makes-ultra-strong-body-armour.

37. Victor C. Li, "Bendable Concrete, with a Design Inspired by Seashells, Can Make US Infrastructure Safer and More Durable," The Conversation, May 25, 2018, https://theconversation.com/bendable-concrete-with-a-design-inspired-by-seashells-can-make-us-infrastructure-safer-and-more-durable-93621.

38. "What Is Hempcrete?" American Lime Technology, http://www.americanlimetechnology.com/what-is-hempcrete.

39. "Seven New Materials Could Change Buildings," *Dino's Storage* (blog), http://dinosstorage.com/wordpress/?p=1150.

40. Wikipedia, s.v. "Metamaterial," last modified July 18, 2019, 17:01, https://en.wikipedia.org/wiki/Metamaterial.

41. Ali Morris, "Dutch Designers Convert Algae into Bioplastic for 3D Printing," Dezeen, December 4, 2017, https://www.dezeen.com/2017/12/04/dutch-designers-eric-klarenbeek-maartje-dros-convert-algae-biopolymer-3d-printing-good-design-bad-world.

42. M. K. Enamala et al., "Production of Biofuels from Microalgae," *Renewable and Sustainable Energy Reviews* 94 (October 2018): 49–68, https://doi.org/10.1016/j.rser.2018.05.012.

43. Kimberley Mok, "Newspapers Recycled into Paper Timber & Furniture by Mieke Meijer," Treehugger, July 12, 2011, https://www.treehugger.com/green-architecture/newspapers-recycled-into-paper-timber-furniture-by-mieke-meijer-vij5.html.

44. "The Origami Revolution," *NOVA*, PBS, https://www.pbs.org/wgbh/nova/video/the-origami-revolution.

45. "Polymer Paper," Alibaba, https://www.alibaba.com/showroom/polymer-paper.html.

46. Emma Betuel, "Scientists Discover Bizarre Way to Build a House Out of Human Urine," *Inverse*, October 27, 2018, https://www.inverse.com/article/50238-sustainable-bio-bricks-made-of-human-urine.

47. "Plaited Microbial Cellulose," Building Centre, https://www.buildingcentre.co.uk/supermaterial/plaited-microbial-cellulose.

48. Wikipedia, s.v. "Bacterial Cellulose," last modified August 20, 2019, 20:48, https://en.wikipedia.org/wiki/Bacterial_cellulose.

49. Kelsey Campbell-Dollaghan, "7 New Materials That Could Change How Our Buildings Act," Gizmodo, August 25, 2014, https://gizmodo.com/6-supermaterials-that-could-change-how-our-buildings-ac-1626274391.

50. Kirsty Sier, "An Ode to Bio-Receptive Concrete," *Architecture & Design*, April 27, 2017, https://www.architectureanddesign.com.au/features/product-in-focus/an-ode-to-bio-receptive-concrete.

51. Jeremy Deaton, "Wood-and-Glue Skyscrapers Are on the Rise," *Popular Science*, April 26, 2016, https://www.popsci.com/wood-and-glue-skyscrapers-are-on-rise.

52. Katrina Filippidis, "Researchers Build a Self-Healing 'Robot Skin,'" Engadget, May 28, 2018, https://www.engadget.com/2018/05/28/self-healing-robot-skin.

53. Ben Brumfield, "New Fire Extinguisher: Bass Hum Booms Flames Out," CNN, March 30, 2015, https://www.cnn.com/2015/03/27/us/sound-fire-extinguisher/index.html.

54. Kristin Houser, "Microorganisms That Eat Seaweed Can Create Biodegradable Plastic," Futurism, December 28, 2018, https://futurism.com/the-byte/bioplastic-microorganisms-eat-seaweed.

55. Cheyenne MacDonald, "'Super Concrete' Made with Fungus Can Heal Itself If It Cracks," *Daily Mail Online*, January 18, 2018, https://www.dailymail.co.uk/sciencetech/article-5285447/Fungus-pave-way-self-healing-concrete.html.

56. David Greig, "Self-Healing Car Paint Uses Sunlight to Repair Scrapes," New Atlas, March 18, 2009, https://newatlas.com/self-healing-car-paint/11254.

57. Wikipedia, s.v. "Metal Foam."

58. Nick Heath, "From Self-Assembling Furniture to Curving Racing Car Wings, Seven GIFs That Show the Future of 4D Printing," Tech Republic, May 1, 2015, https://www.techrepublic.com/pictures/from-self-assembling-furniture-to-curving-racing-car-wings-seven-gifs-that-show-the-future-of-4d.

59. "Shrilk Biodegradable Plastic," Wyss Institute, https://wyss.harvard.edu/technology/chitosan-bioplastic.

60. Princeton University, Engineering School, "Self-Powered System Makes Smart Windows Smarter," Science Daily, June 30, 2017, https://www.sciencedaily.com/releases/2017/06/170630115619.htm.

61. Kevin Ohannessian, "Transparent Solar Panels May Be a Window into the Future," *Tech Times*, August 20, 2014, https://www.techtimes.com/articles/13575/20140820/transparent-solar-panel-cell-michigan-state-university.htm.

62. "New Paint Additive Turns Any Surface into a Solar Panel," ProudGreenHome.com, July 8, 2015, https://www.proudgreenhome.com/news/new-paint-additive-turns-any-surface-into-a-solar-panel.

63. "Solar Shingles," *Solar Energy Facts* (blog), http://solarenergyfactsblog.com/solar-shingles.

64. "What Is Soil Stabilizing?" AggreBind, https://aggrebind.com/about/what-is-a-polymer-what-is-soil-stabilizing.

65. James Vincent, "New Wonder Material 'Stanene' Could Replace Graphene with 100% Electrical Conductivity," *The Independent*, https://www.independent.co.uk/news/science/new-wonder-material-stanene-could-replace-graphene-with-100-electrical-conductivity-8967573.html.

66. "Structural Insulated Panels," EPS Industry Alliance, https://www.epsindustry.org/building-construction/structural-insulated-panels.

67. Aisha Abdelhamid, "Bamboo: The ABCs of Green Building Materials," Green Building Elements, August 10, 2016, https://greenbuildingelements. com/2016/08/10/bamboo-abcs-green-building-materials.

68. Jon Evans, "Nanotech Clothing Fabric 'Never Gets Wet,'" *New Scientist*, November 24, 2008, https://www.newscientist.com/article/dn16126-nanotech-clothing-fabric-never-gets-wet.

69. Erik North, "What Is Thermal Bridging?" GreenBuildingAdvisor, April 15, 2013, https://www.greenbuildingadvisor.com/article/what-is-thermal-bridging.

70. "Item #: 72-TCD-.125M Thermally Conductive," Thomas, https:// certifications.thomasnet.com/catalogs/item/10110251-1472-3001106-2786/cs-hyde-company/thermally-conductive.

71. *TIMBERCRETE®: An Introduction*, Timbercrete, http://www. timbercrete.com.au/pdfs/Introduction_to_Timbercrete.pdf.

72. G. P. Thomas, "Transparent Aluminum (Aluminum Oxynitride): Properties, Production and Applications," Azo Materials, February 1, 2013, https:// www.azom.com/article.aspx?ArticleID=8095.

73. US Department of Energy, *Breaking the Biological Barriers to Cellulosic Ethanol*, June 2006, https://genomicscience.energy.gov/biofuels/ 2005workshop/b2bhighres63006.pdf.

74. Juan Rodriguez, "Why to Consider Triple Glazed Windows for Your Home," The Balance: Small Business, last modified June 25, 2019, https:// www.thebalancesmb.com/triple-glazed-windows-844733.

75. Wikipedia, s.v. "Vantablack," last modified August 20, 2019, 15:28, https://en.wikipedia.org/wiki/Vantablack.

76. Bjorn Peter Jelle, "Vacuum Insulation Panel Products: A State-of-the-Art Review and Future Research Pathways," *Applied Energy* 116 (March 2014): 355–75.

77. Rick Duncan, "Benefits of SPF in Insulation Applications," *Construction Specifier*, January 9, 2015, https://www.constructionspecifier.com/the-benefits-of-spf-in-insulation-applications.

78. "Innovations in Construction," *Blog*, Bautex, https://www. bautexsystems.com/BLOG/INNOVATIONS-IN-CONSTRUCTION.

79. "Seaweed, Wool Make Traditional Bricks Tougher," *This Just In* (blog) CNN, October 5, 2010, http://news.blogs.cnn.com/2010/10/05/seaweed-wool-make-traditional-bricks-tougher.

8. GETTING SOMEWHERE
FROM SOMEPLACE

1. *The Road to Sustainable Urban Logistics*, UPS, 2017, https:// sustainability.ups.com/media/UPS_The_Road_to_Sustainable_Urban_ Logistics.pdf.

2. "Sources of Greenhouse Gas Emissions," US EPA, https://www.epa.gov/ ghgemissions/sources-greenhouse-gas-emissions.

3. Sarah McDermott, "Who's Driving This Bus? Nobody," CNET, April 29, 2017, https://www.cnet.com/roadshow/news/self-driving-cars-automated-public-transport-bus.

4. Amanda Cunningham, "Public Transportation of the Future: Four New Sustainable Technologies," *HOK* (blog), *Building Design + Construction*, January 31, 2017. https://www.bdcnetwork.com/blog/public-transportation-future-four-new-sustainable-technologies.

5. Cunningham, "Public Transportation of the Future."

6. Cunningham, "Public Transportation of the Future."

7. Jared Ficklin, "Future-Proofing Transportation: The Missing Opportunity for Our Cities," NewCities, https://newcities.org/perspectives-future-proofing-transportation-the-missing-opportunity-for-our-cities.

8. Susan Fourtané, "Connected Vehicles in Smart Cities: The Future of Transportation," Interesting Engineering, November 16, 2018, https://interestingengineering.com/connected-vehicles-in-smart-cities-the-future-of-transportation.

9. Michelle Davidson, "Smart City Challenge: 7 Proposals for the Future of Transportation," Network World, June 16, 2016, https://www.networkworld.com/article/3084455/smart-city-challenge-7-proposals-for-the-future-of-transportation.html.

10. "New Data Shows 94 Percent of Car Accidents Caused by Human Error," *Georgia Personal Injury Blog*, Southside Injury Attorneys, July 21, 2016, https://southsideinjuryattorneys.com/lawyer/2016/07/21/Personal-Injury/New-Data-Shows-94-Percent-of-Car-Accidents-Caused-by-Human-Error_bl25860.htm.

11. Centers for Disease Control and Prevention, National Center for Injury Prevention and Control, "State-Specific Costs of Motor Vehicle Crash Deaths," CDC, last modified December 14, 2015, https://www.cdc.gov/motorvehiclesafety/statecosts/index.html.

12. "Many Drivers Spend 127 Hours a Year in Traffic," Sky News, March 22, 2016, https://news.sky.com/story/many-drivers-spend-127-hours-a-year-in-traffic-10214621.

13. Angie Schmitt, "It's True: The Typical Car Is Parked 95 Percent of the Time," *Streetsblog USA*, March 10, 2016, https://usa.streetsblog.org/2016/03/10/its-true-the-typical-car-is-parked-95-percent-of-the-time.

14. Todd Neff, "A Trillion Bucks Says You'll Sell Your Wheels," *Solutions Journal*, June 17, 2015, https://medium.com/solutions-journal-summer-2015/a-trillion-bucks-says-you-ll-sell-your-wheels-d4562be9d3d7.

15. "Predictions on the Increase in Cross-Border Commuters over the Next 40 Years," Grand Duchy of Luxembourg, May 16, 2017, http://luxembourg.public.lu/en/actualites/2017/05/16-frontaliers/index.html.

16. Kristin Houser, "Luxembourg Just Made Public Transportation Free for Everyone," Futurism, December 6, 2018, https://futurism.com/luxembourg-public-transportation-free.

17. Jeff Wattenhofer, "14 Percent of Los Angeles County Land Is Dedicated to Parking," Curbed, November 30, 2015, https://la.curbed.com/2015/11/30/9895842/how-much-parking-los-angeles.

18. Brad Plumer, "Cars Take Up Way Too Much Space in Cities," Vox, https://www.vox.com/a/new-economy-future/cars-cities-technologies.

19. Melissa Wylie, "The Cost of Parking May Surprise You," *Biz Women*, April 12, 2018, https://www.bizjournals.com/bizwomen/news/latest-news/2018/04/the-cost-of-parking-may-surprise-you.html.

20. Adele Peters, "See Just How Much of a City's Land Is Used for Parking Spaces," *Fast Company*, July 20, 2017, https://www.fastcompany.com/40441392/see-just-how-much-of-a-citys-land-is-used-for-parking-spaces.

21. Dom Galeon, "Toyota's New Self-Driving Cars Will Chat with Drivers," NBC News, October 18, 2017, https://www.nbcnews.com/mach/science/toyota-s-new-self-driving-cars-will-chat-drivers-ncna811826.

22. Karla Lant, "Get Ready for Self-Driving Cars to Hit the Road," Futurism, March 26, 2017, https://futurism.com/get-ready-self-driving-cars-hit-road.

23. Aarian Marshall, "Elon Musk Unveils the Boring Company's Car-Flinging Tunnel," *Wired*, December 19, 2018, https://www.wired.com/story/elon-musk-boring-company-car-flinging-tunnel; Ed Brackett, "Elon Musk: Rapid-Transit Test Tunnel under LA Opens to Public Dec. 10," CNBC, October 22, 2018, https://www.cnbc.com/2018/10/22/elon-musk-rapid-transit-test-tunnel-under-la-opens-to-public-dec-10.html.

24. Dan Robitzski, "Cheap Hydrogen Fuel Was a Failed Promise. But Its Time May Have Arrived," Futurism, April 30, 2018, https://futurism.com/ammonia-hydrogen-fuel.

25. Robitzski, "Cheap Hydrogen Fuel Was a Failed Promise."

26. Liane Yvkoff, "Is a Nuclear-Powered Car in Our Future?" Roadshow, September 1, 2011, https://www.cnet.com/roadshow/news/is-a-nuclear-powered-car-in-our-future.

27. Emma Taggart, "IKEA's First India Store Makes Deliveries with Colorful Solar-Powered Rickshaws," My Modern Met, August 13, 2018, https://mymodernmet.com/ikea-india-rickshaws; "Solar Electric Rickshaw," Alibaba, https://www.alibaba.com/showroom/solar-electric-rickshaw.html.

28. June Javelosa, "The World's First Flying Taxis Will Take to the Skies in Five Months," Futurism, February 16, 2017, https://futurism.com/the-worlds-first-flying-taxis-will-take-to-the-skies-in-five-months.

29. Dom Galeon, "Report Hints Porsche Might Have a Passenger Drone in the Works," Futurism, March 4, 2018, https://futurism.com/porsche-passenger-drone.

30. Kyree Leary, "AirSpaceX's Autonomous Electric Flying Taxi Will Hit the Skies in 2026," Futurism, January 26, 2018, https://futurism.com/airspacex-autonomous-electric-flying-taxi-hit-skies-2026.

31. "Mercedes' Parent Company Makes a Big Investment in a Flying Taxi Start-Up: Here's What We Know," *Style*, August 14, 2017, https://www.scmp.com/magazines/style/news-trends/article/2106676/mercedes-parent-company-makes-big-investment-flying-taxi.

32. "Passenger Drones Are a Better Kind of Flying Car," *The Economist*, March 10, 2018, https://www.economist.com/science-and-technology/2018/03/10/passenger-drones-are-a-better-kind-of-flying-car.

33. Stacy Liberatore and Mark Prigg, "Is This the Future of Commuting? Watch the First Manned Flight of the Volocopter 'Personal Drone' with 18 Rotors," *Daily Mail Online*, April 7, 2016, https://www.dailymail.co.uk/sciencetech/article-3528834/Watch-manned-flight-personal-drone-18-rotors-Volocopter-replace-car-flies-passenger-groundbreaking-test.html.

34. Tony Aube, "Our Self-Flying Car Future," TechCrunch, December 23, 2016, https://techcrunch.com/2016/12/23/our-self-flying-car-future.

35. Christine Negroni, "Before Flying Car Can Take Off, There's a Checklist," *New York Times*, April 27, 2012, https://www.nytimes.com/2012/04/29/automobiles/before-flying-car-can-take-off-theres-a-checklist.html.

36. Victor Tangermann, "Uber Gave Architects the Chance to Design Their Sci-Fi Flying Taxi 'Skyports,' and They Came Back with This," Futurism, May 11, 2018, https://futurism.com/uber-flying-taxi-skyports-concepts.

37. Dan Robitzski, "Uber Plans to Launch Flying Taxis with Technology That Doesn't Exist," Futurism, May 7, 2018, https://futurism.com/uber-flying-taxis.

38. Victor Tangermann, "Elon Musk Says an Upcoming Tesla Roadster Will Hover: Is He Joking?" Futurism, January 10, 2019, https://futurism.com/elon-musk-new-tesla-roadster-hover-spacex.

39. Teo Kermeliotis, "Solar-Powered Roads: Coming to a Highway Near You?" CNN, September 18, 2014, http://www.cnn.com/2014/05/12/tech/solar-powered-roads-coming-highway/index.html.

40. "Overview," Solar Roadways, http://www.solarroadways.com/Specifics/Solar; Mike Murphy, "The World's First Solar Panel-Paved Road Has Opened in France," Quartz, December 23, 2016, https://qz.com/871162/the-first-road-paved-in-solar-panels-opened-in-france.

41. "ELF 2FR," https://organictransit.com/product/elf-2fr.

42. "Organic Transit," https://organictransit.com; Valerie Bonk, "Not a Car, not a Bike, but a Blend: an ELF vehicle," *Lowell Sun Online*, last modified August 1, 2013, http://www.lowellsun.com/ci_23774175/not-car-not-bike-but-blend-an-elf.

43. Tom DiChristopher, "Electric Vehicles Will Grow from 3 Million to 125 Million by 2030, International Energy Agency Forecasts," CNBC, May 30, 2018, https://www.cnbc.com/2018/05/30/electric-vehicles-will-grow-from-3-million-to-125-million-by-2030-iea.html.

44. Jimmy O'Dea, "Electric vs. Diesel vs. Natural Gas: Which Bus Is Best for the Climate?" *Union of Concerned Scientists* (blog), July 19, 2018, https://blog.ucsusa.org/jimmy-odea/electric-vs-diesel-vs-natural-gas-which-bus-is-best-for-the-climate.

45. "Elon Musk's Ideas about Transportation are Boring," *Pedestrian Observations* (blog), December 15, 2017, https://pedestrianobservations.com/2017/12/15/elon-musks-ideas-about-transportation-are-boring.

46. Joseph Flynt, "Choosing the Right Sized Drone," 3D Insider, September 12, 2017, https://3dinsider.com/drone-sizes.

47. Annie Sneed, "So Your Neighbor Got a Drone for Christmas," *Scientific American*, December 22, 2015, https://www.scientificamerican.com/article/so-your-neighbor-got-a-drone-for-christmas.

48. Victor Tangermann, "Dubai Police Are Training Officers to Fly Hoverbikes," Futurism, November 8, 2018, https://futurism.com/dubai-police-training-officers-hoverbikes.

49. "Make a Hovercraft Powered by an Electric Leaf Blower," Gadget Hacks, January 30, 2008, https://mods-n-hacks.gadgethacks.com/how-to/make-hovercraft-powered-by-electric-leaf-blower-127038.

50. Sebastian Anthony, "For $10,000, You Can Have the World's First Hoverboard," Extreme Tech, October 21, 2014, https://www.extremetech.com/extreme/192508-for-10000-you-can-have-the-worlds-first-hoverboard-and-invest-in-earthquake-proof-levitating-homes.

51. Sean O'Kane, "Yes, the Jet-Powered Hoverboard Is Real, and Yes, the Creator Has Crashed It," The Verge, April 15, 2016, https://www.theverge.com/2016/4/15/11439798/franky-zapata-racing-jet-powered-flying-hoverboard-interview; Eric Adams, "Real-Life Flying Suit Inventor Richard Browning to Launch Jetpack Racing League Next Year," The Drive, September 10, 2018, https://www.thedrive.com/tech/23483/real-life-flying-suit-inventor-richard-browning-to-launch-gravity-jetpack-racing-series-in-2019.

52. Alex Hutchinson, "Jet Pack for Sale: Only $100,000!" *Popular Mechanics*, July 9, 2010, https://www.popularmechanics.com/flight/how-to/a5892/martin-aircraft-jet-pack-for-sale.

53. Ed Grabianowski, "How Jet Packs Work," HowStuffWorks, July 7, 2008, https://science.howstuffworks.com/transport/engines-equipment/jet-pack3.htm.

54. "200mph JB11 JetPack to Make European Flight Debut at Goodwood Festival of Speed," Racecar, June 5, 2018, https://www.racecar.com/News/88815/Motorsport/200mph-JB11-JetPack.

55. Adams, "Real-Life Flying Suit Inventor Richard Browning to Launch Jetpack Racing League Next Year."

56. Jack Nicas, "Is the Jetpack Movement Finally Taking Off?" *Wall Street Journal*, June 7, 2016, https://www.wsj.com/articles/is-the-jetpack-movement-finally-taking-off-1465221130.

57. Grabianowski, "How Jet Packs Work."

58. "Hypersonic Flight," National Air and Space Museum, May 30, 2012, https://airandspace.si.edu/stories/editorial/hypersonic-flight.

59. Kyree Leary, "China's New Hypersonic Plane Travels from Beijing to New York in a Few Hours," Futurism, February 26, 2018, https://futurism.com/china-hypersonic-plane.

60. Ankit Ajmera, "Japan Airlines Invests $10 Million in Supersonic Jet Company Boom," Reuters, December 5, 2017, https://www.reuters.com/article/us-boom-japan-airlines/japan-airlines-invests-10-million-in-supersonic-jet-company-boom-idUSKBN1DZ1N2; National Research Council, "Conclu-

sions and Policy Recommendations," in *High-Stakes Aviation: U.S.-Japan Technology Linkages in Transport Aircraft* (Washington, DC: National Academies Press, 1994), https://www.nap.edu/read/2346/chapter/7.

61. Matt Williams, "How Fast Is Mach One?" Universe Today, July 11, 2017, https://www.universetoday.com/77077/how-fast-is-mach-1; Wikipedia, s.v. "Supersonic Speed," last modified July 15, 2019, 19:35, https://en.wikipedia.org/wiki/Supersonic_speed.

62. Eric M. Johnson, "Paul Allen's Space Firm Details Plans for Rockets, Cargo Vehicle," Reuters, August 20, 2018, https://www.reuters.com/article/us-space-paulallen/paul-allens-space-firm-details-plans-for-rockets-cargo-vehicle-idUSKCN1L52AY.

63. "Pegasus," Northrop Grumman, https://www.northropgrumman.com/Capabilities/Pegasus/Pages/default.aspx.

64. Jeff Foust, "Pegasus Woes Continue to Delay NASA Mission," *Space News*, February 28, 2019, https://spacenews.com/pegasus-woes-continue-to-delay-nasa-mission.

65. Jeff Foust, "How High Is Space?" Space Review, August 10, 2009, http://www.thespacereview.com/article/1436/1.

66. Wikipedia, s.v. "Space Tourism," last modified July 22, 2019, 14:28, https ://en.wikipedia.org/wiki/Space_tourism; John Antczak, "Virgin Galactic Tourism Rocket Ship Reaches Space in Test," AP News, December 13, 2018, https://www.apnews.com/659f385710cc46fdb381c5f6dfbb6573.

67. Adam Mann, "So You Want to Be a Space Tourist? Here Are Your Options," NBC News, July 21, 2017, https://www.nbcnews.com/mach/science/so-you-want-be-space-tourist-here-are-your-options-ncna784166.

68. Mike Wall, "First Space Tourist: How a U.S. Millionaire Bought a Ticket to Orbit," Space, https://www.space.com/11492-space-tourism-pioneer-dennis-tito.html.

69. "Elon Musk Unveils First Tourist for SpaceX 'Moon Loop,'" BBC News, September 18, 2018, https://www.bbc.com/news/science-environment-45550755.

70. Matt Williams, "Aerojet Rocketdyne Tests Out Its New Advanced Ion Engine System," Universe Today, September 3, 2018, https://www.universetoday.com/tag/ion-engine.

71. "iLint: The World's First Hydrogen-Powered Train," Railway Technology, January 22, 2018, https://www.railway-technology.com/features/ilint-worlds-first-hydrogen-powered-train.

72. Kevin Bonsor and Nathan Chandler, "How Maglev Trains Work," How-StuffWorks, https://science.howstuffworks.com/transport/engines-equipment/maglev-train.htm; Wikipedia, s.v. "Magnetic Levitation."

73. "Working of MAGLEV Trains," Circuits Today, February 16, 2012, http://www.circuitstoday.com/working-of-maglev-trains.

74. James Glave and Rachel Swaby, "Superfast Bullet Trains Are Finally Coming to the U.S.," *Wired*, January 25, 2010, https://www.wired.com/2010/01/ff_fasttrack.

75. Danny Paez, "Elon Musk Says 'Sure,' the Boring Company Can Transform This City's Transit," Inverse, November 6, 2018, https://www.inverse.com/article/50574-elon-musk-the-boring-company-new-city.

76. Arjun Kharpal, "Elon Musk's Hyperloop Vision Takes a Step Closer to Reality as Firm Reveals Pictures of Test Track," CNBC, March 9, 2017, https://www.cnbc.com/2017/03/09/elon-musk-hyperloop-closer-to-reality-test-track-nevada.html.

77. "Is Hyperloop the Future of Transportation," *CIO Review*, July 7, 2016, https://www.cioreview.com/news/is-hyperloop-the-future-of-transportation-nid-15226-cid-89.html.

78. Natalie Burkhard, "Why Invent the Hyperloop?" Stanford University, December 11, 2014, http://large.stanford.edu/courses/2014/ph240/burkhard2.

79. Denise Chow, "This Robotic Manta Ray May Speed Underwater Search and Rescue," NBC News, December 8, 2017, https://www.nbcnews.com/mach/science/robotic-manta-ray-may-speed-underwater-search-rescue-ncna827806.

80. Kastalia Medrano, "5 Robot Animals You Need to Know About," Inverse, April 4, 2016, https://www.inverse.com/article/13741-5-robot-animals-you-need-to-know-about.

81. Jude Garvey, "Innespace's Seabreacher: Cross between a Dolphin and a PWC," New Atlas, September 25, 2009, https://newatlas.com/innespaces-seabreacher/12943.

9. PRIORITIES FOR POWER

1. Stephen Leahy, "Cities Emit 70% More Carbon Than Thought," *National Geographic*, March 6, 2018, https://news.nationalgeographic.com/2018/03/city-consumption-greenhouse-gases-carbon-c40-spd.

2. World Energy Council, *World Energy Resources: 2016* (London: World Energy Council, October 2016), https://www.worldenergy.org/wp-content/uploads/2016/10/World-Energy-Resources_SummaryReport_2016.pdf.

3. Bobby Magill, "Americans Used a Lot Less Coal in 2016," *Scientific American*, April 7, 2017, https://www.scientificamerican.com/article/americans-used-a-lot-less-coal-in-2016.

4. Jeff Brady, "Trump Administration Weakens Climate Plan to Help Coal Plants Stay Open," NPR, June 19, 2019, https://www.npr.org/2019/06/19/733800856/trump-administration-weakens-climate-plan-to-help-coal-plants-stay-open.

5. US Energy Information Administration, "Natural Gas Expected to Surpass Coal in Mix of Fuel Used for U.S. Power in 2016," EIA, March 16, 2016, https://www.eia.gov/todayinenergy/detail.php?id=25392.

6. US EPA, *Hydraulic Fracturing for Oil and Gas: Impacts from the Hydraulic Fracturing Water Cycle on Drinking Water Resources in the United States*, Final Report (Washington, DC: US Environmental Protection Agency, 2016), https://cfpub.epa.gov/ncea/hfstudy/recordisplay.cfm?deid=332990.

7. Laignee Barron, "Here's What the EPA's Website Looks Like after a Year of Climate Change Censorship," *Time*, March 1, 2018, https://time.com/5075265/epa-website-climate-change-censorship; Brady Dennis and Juliet Eilperin, "Trump Signs Order at the EPA to Dismantle Environmental Protections," *Washington Post*, March 28, 2017, https://www.washingtonpost.com/national/health-science/trump-signs-order-at-the-epa-to-dismantle-environmental-protections/2017/03/28/3ec30240-13e2-11e7-ada0-1489b735b3a3_story.html.

8. "Stanford Engineers Develop State-by-State Plan to Convert U.S. to 100% Clean, Renewable Energy by 2050," *Stanford News*, June 8, 2015, https://news.stanford.edu/2015/06/08/50states-renewable-energy-060815.

9. A. García-Olivares, J. Solé, and O. Osychenko, "Transportation in a 100% Renewable Energy System," Energy Conversion and Management 158 (February 15, 2018): 266–85, https://www.sciencedirect.com/science/article/pii/S0196890417312050.

10. David Coady et al., "Global Fossil Fuel Subsidies Remain Large: An Update Based on Country-Level Estimates," IMF Working Papers, May 2, 2019, https://www.imf.org/en/Publications/WP/Issues/2019/05/02/Global-Fossil-Fuel-Subsidies-Remain-Large-An-Update-Based-on-Country-Level-Estimates-46509.

11. Meyer Robinson, "The Hidden Subsidy of Fossil Fuels," The Atlantic, May 9, 2019, https://www.theatlantic.com/science/archive/2019/05/how-much-does-world-subsidize-oil-coal-and-gas/589000.

12. US Department of Energy, "Oil Shale and Other Unconventional Fuels Activities," Energy.gov, https://www.energy.gov/fe/services/petroleum-reserves/naval-petroleum-reserves/oil-shale-and-other-unconventional-fuels.

13. "Global Demand for Energy Will Peak in 2030, Says World Energy Council," *The Guardian*, October 10, 2016, https://www.theguardian.com/business/2016/oct/10/global-demand-for-energy-will-peak-in-2030-says-world-energy-council.

14. J. de Boer and C. Zuidema, "Towards an Integrated Energy Landscape," University of Groningen, 2013, https://www.rug.nl/research/portal/publications/towards-an-integrated-energy-landscape(dc6d4030-eea1-4e97-ac5e-2180b575f568)/export.html.

15. David Worthington, "Empire State Building Becomes Model for Energy Efficiency," ZDNet, June 7, 2012, https://www.zdnet.com/article/empire-state-building-becomes-model-for-energy-efficiency.

16. Wikipedia, s.v. "Ocean Thermal Energy Conversion," last modified July 18, 2019, 22:34, https://en.wikipedia.org/wiki/Ocean_thermal_energy_conversion.

17. "Hydrogen from Waste Materials," Alternative Energy News, https://www.alternative-energy-news.info/hydrogen-from-waste-materials.

18. Christina Nunez, "Biofuels, Explained," *National Geographic*, July 15, 2019, https://www.nationalgeographic.com/environment/global-warming/biofuel.

19. Wagdy Sawahel, "Chicken Waste Makes Cheap, Food-Friendly Bio-fuel," SciDevNet, August 12, 2009, https://www.scidev.net/global/biofuels/news/chicken-waste-makes-cheap-food-friendly-biofuel-.html; American Chemical Society, "Biodiesel on the Wing: A 'Green' Process for Biodiesel from Feather Meal," ScienceDaily, July 23, 2009, https://www.sciencedaily.com/releases/2009/07/090722110903.htm.

20. Nick Bilton, "Wireless Charging, at a Distance, Moves Forward for uBeam," *Bits* (blog), *New York Times*, August 6, 2014, https://bits.blogs.nytimes.com/2014/08/06/ubeam-technology-will-enable-people-to-charge-devices-through-the-air.

21. "Inductive versus Resonant Wireless Charging," Digi-Key, August 2, 2016, https://www.digikey.com/en/articles/techzone/2016/aug/inductive-versus-resonant-wireless-charging.

22. Adele Peters, "Can These 35-Ton Bricks Solve Renewable Energy's Biggest Problem?" *Fast Company*, November 7, 2018, https://www.fastcompany.com/90261233/can-these-35-ton-bricks-solve-renewable-energys-biggest-problem.

23. Jeff Dondero, *Throwaway Nation: The Ugly Truth about American Garbage* (Lanham, MD: Rowman & Littlefield, 2019).

24. Camila Domonoske, "California Sets Goal of 100 Percent Clean Electric Power by 2045," NPR, September 10, 2018, https://www.npr.org/2018/09/10/646373423/california-sets-goal-of-100-percent-renewable-electric-power-by-2045.

25. "State Renewable Portfolio Standards and Goals," NCSL, http://www.ncsl.org/research/energy/renewable-portfolio-standards.aspx; "Tracking Progress," California Energy Commission, https://www.energy.ca.gov/data-reports/tracking-progress.

26. "The Spherical Sun Power Generator," Alternative Energy News, April 13, 2018, http://www.alternative-energy-news.info/spherical-sun-power-generator; Matt Klassen, "Spherical Sun Power Generator," Stanford University, September 14, 2017, http://large.stanford.edu/courses/2016/ph240/klassen2.

27. Patrick Caughill, "A New 'Solar Paint' Lets You Transform Your Entire House into a Source of Clean Energy," Futurism, June 15, 2017, https://futurism.com/a-new-solar-paint-lets-you-transform-your-entire-house-into-a-source-of-clean-energy.

28. Caughill, "A New 'Solar Paint' Lets You Transform Your Entire House into a Source of Clean Energy."

29. "ANU Researchers Use Sticky Tape to Create Ultra-Thin Film Solar Cell Material," Alternative Energy, http://www.altenergy.org/renewables/ultrathin-solarcells.html.

30. "Development Begins on World's First Electricity-Generating Windows for Retrofitting Existing Homes and Commercial Buildings," Solar Window Technologies, September 6, 2016, https://www.solarwindow.com/2016/09/development-begins-worlds-first-electricity-generating-windows-retrofitting-existing-homes-commercial-buildings; Catharine Paddock, "Electricity-Gener-

ating Veneers Will Turn Windows into Solar Panels," Market Business News, September 13, 2016, https://marketbusinessnews.com/electricity-generating-veneers-will-turn-windows-solar-panels/144223.

31. B. J. Trześniewski and W. A. Smith, "Photocharged $BiVO_4$ Photoanodes for Improved Solar Water Splitting," *Journal of Materials Chemistry A* 4, no. 8 (2016), https://pubs.rsc.org/en/content/articlelanding/2016/ta/c5ta04716a#!divAbstract.

32. Nidhi Goyal, "KBNNO Is a New Material That Can Turn Sunlight, Heat, and Movement into Electricity," Industry Tap, February 27, 2017, http://www.industrytap.com/kbnno-new-material-can-turn-sunlight-heat-movement-electricity/41068.

33. Q. Chen et al., "Under the Spotlight: The Organic–Inorganic Hybrid Halide Perovskite for Optoelectronic Applications," *NanoToday* 10, no. 3 (June 2015): 355–96, https://www.sciencedirect.com/science/article/pii/S1748013215000560.

34. Goyal, "KBNNO Is a New Material That Can Turn Sunlight, Heat, and Movement into Electricity."

35. Charles Choi, "Out-of-This-World Proposal for Solar Wind Power," *New Scientist*, September 24, 2010, https://www.newscientist.com/article/dn19497-out-of-this-world-proposal-for-solar-wind-power.

36. Bob Vila, "Discover Ideas about Garden Paths," Pinterest, https://www.pinterest.com/pin/270075308879723906; "Solar-Powered Sun Brick," Inhabitat, November 6, 2006, https://inhabitat.com/solar-powered-sun-brick/sun-brick-2.

37. Wikipedia, s.v. "Solar Tree," last modified June 10, 2018, 20:58, https://en.wikipedia.org/wiki/Solar_tree; "Gardens by the Bay: Super Trees Generate Solar Power," Go Green, https://www.go-green.ae/greenstory_view.php?storyid=2085.

38. "Hydrelio® Floating Solar's Benefits," Ciel & Terre, https://www.ciel-et-terre.net/hydrelio-floating-solar-technology/hydrelio-benefits.

39. "MIT Breakthrough: Thermo-Chemical Solar Power," Alternative Energy News, November 3, 2010, http://www.alternative-energy-news.info/mit-thermo-chemical-solar-power.

40. Dan Robitzski, "A Purple, Photosynthetic Bacteria Can Turn Your Poop into Power," Futurism, November 13, 2018, https://futurism.com/the-byte/purple-photosynthetic-bacteria-poop-power.

41. Kristin Houser, "A Tiny Crustacean Could Help Us Create Biofuel from Wood," Futurism, December 4, 2018, https://futurism.com/the-byte/biofuel-wood-crustacean-gribble.

42. "Nanotube Technology Transforms CO_2 into Fuel," Alternative Energy News, https://www.alternative-energy-news.info/nanotube-technology-transforms-co2-into-fuel.

43. "Pyromex Waste to Energy Technology," Alternative Energy News, https://www.alternative-energy-news.info/pyromex-waste-energy.

44. US Department of Energy, "Hydrogen Production: Electrolysis," https://www.energy.gov/eere/fuelcells/hydrogen-production-electrolysis.

45. Rebecca Boyle, "Nanotube-Tethered Flying Wind Turbines Could Harvest Energy at 30,000 Feet," *Popular Science*, December 17, 2010, https://www.popsci.com/technology/article/2010-12/nano-tethered-flying-wind-turbines-inspire-new-nasa-study.

46. "New Design Gets the Heat Out of Fusion Reactors," New Energy and Fuel, October 16, 2018, https://newenergyandfuel.com/2018/10/16/new-design-gets-the-heat-out-of-fusion-reactors.

47. "Nuclear Fusion Power," World Nuclear Association, updated February 2019, https://www.world-nuclear.org/information-library/current-and-future-generation/nuclear-fusion-power.aspx; Hannah Devlin, "Nuclear Fusion on Brink of Being Realized," *The Guardian*, March 9, 2018, https://www.theguardian.com/environment/2018/mar/09/nuclear-fusion-on-brink-of-being-realised-say-mit-scientists.

48. "The Man Putting Coffee into Fuel," Shell Global, https://www.shell.com/inside-energy/coffee-into-fuel.html; "Turning Coffee Grounds into Biofuel Is More Efficient Than Ever," Curiosity, May 25, 2017, https://curiosity.com/topics/turning-coffee-grounds-into-biofuel-is-more-efficient-than-ever-curiosity.

49. Jim Motavalli, "8 Alternative Fuels That Could Replace Oil," *Popular Mechanics*, May 4, 2010, https://www.popularmechanics.com/cars/hybrid-electric/g58/alternative-fuel-cars-460509.

50. Killian Fox, "The Floor Tiles That Use Foot Power to Light Up Cities," *The Guardian*, January 11, 2015, https://www.theguardian.com/technology/2015/jan/11/floor-tile-generates-power-from-footsteps-energy-electricity-startup.

51. Debbie Sniderman, "Energy Efficient Elevator Technologies," ASME, September 19, 2012, https://www.asme.org/topics-resources/content/energy-efficient-elevator-technologies.

52. GCR Staff, "GE to Build Massive Wind Turbine, Triple the Height of Statue of Liberty," Global Construction Review, January 23, 2019, http://www.globalconstructionreview.com/news/ge-build-massive-wind-turbine-triple-height-statue.

53. Knvul Sheikh, "New Concentrating Solar Tower Is Worth Its Salt with 24/7 Power," *Scientific American*, July 14, 2016, https://www.scientificamerican.com/article/new-concentrating-solar-tower-is-worth-its-salt-with-24-7-power.

54. Wikipedia, s.v. "Hydroelectric Power in the United States," last modified June 16, 2019, 18:58, https://en.wikipedia.org/wiki/Hydroelectric_power_in_the_United_States.

55. "The Pros and Cons of Dams," *Arcadia Power* (blog), https://blog.arcadiapower.com/pros-cons-dams.

56. Dion Lowe, "Harnessing the Power of Nature," SWS, September 3, 2008, https://www.estormwater.com/harnessing-power-nature; Devereaux Bell, "What If We Could Create Energy the Way Nature Does?" MNN, September 15, 2014, https://www.mnn.com/earth-matters/energy/stories/what-if-we-could-create-energy-the-way-nature-does.

57. Payam Adlparvar, "Why Can't We Extract Electricity from Lightning?" *The Independent*, April 11, 2015, https://www.independent.co.uk/news/science/why-cant-we-extract-electricity-from-lightning-10162498.html.

58. "BU-209: How does a Supercapacitor Work?" BatteryUniversity, https://batteryuniversity.com/learn/article/whats_the_role_of_the_supercapacitor.

59. Tom Abate, "Researchers Build a Water-Based Battery to Store Solar and Wind Energy," *Engineers Journal*, June 12, 2018, http://www.engineersjournal.ie/2018/06/12/researchers-build-water-based-battery-store-solar-wind-energy.

60. Rebecca Boyle, "DARPA's Future Li-ion Batteries Will Be Smaller Than Grains of Salt," *Popular Science*, October 20, 2010, https://www.popsci.com/technology/article/2010-10/future-li-ion-batteries-will-be-smaller-grain-salt.

61. Jon Christian, "A New Battery Could Store Ten Times the Power as Lithium-Ion," Futurism, December 9, 2018, https://futurism.com/new-battery-ten-times-power.

62. James Giggacher, "All Power to the Proton: Researchers Make Battery Breakthrough," RMIT, March 8, 2018, https://www.rmit.edu.au/news/all-news/2018/mar/all-power-to-the-proton.

63. "Environmental Damage," Mission 2016: The Future of Strategic Natural Resources, https://web.mit.edu/12.000/www/m2016/finalwebsite/problems/environment.html.

64. Josh Goldman, "Electric Vehicles, Batteries, Cobalt, and Rare Earth Metals," *Union of Concerned Scientists* (blog), October 25, 2017, https://blog.ucsusa.org/josh-goldman/electric-vehicles-batteries-cobalt-and-rare-earth-metals.

65. "Distributed Energy Resources," North America Reliability Corporation, February 2017, https://www.nerc.com/pa/RAPA/ra/Reliability%20Assessments%20DL/Distributed_Energy_Resources_Report.pdf.

66. Jeff Dondero, *The Energy Wise Home* (Lanham, MD: Rowman & Littlefield, 2017).

67. "Buildings of the Future and a More Electric Economy," Fresh Energy, October 9, 2017, https://fresh-energy.org/buildings-of-the-future-and-a-more-electric-economy.

68. Wikipedia, s.v. "Wireless Power Transfer," last modified, June 27, 2019, 19:01, https://en.wikipedia.org/wiki/Wireless_power_transfer; Sourabh Pawade, Tushar Nimje, and Dipti Diwase, "Goodbye Wires: Approach to Wireless Power Transmission," *International Journal of Emerging Technology and Advanced Engineering* 2, no. 4 (April 2012), https://pdfs.semanticscholar.org/39fe/67c7fb77fee2cb50865b7b98aea0c4b2f4fc.pdf.

69. Denise Chow, "A City in China Wants to Launch an Artificial Moon into Space," NBC News, October 25, 2018, https://www.nbcnews.com/mach/science/city-china-wants-launch-artificial-moon-space-ncna923946.

70. Jay Greene, "Trump's Tech Battle with China Roils Bill Gates Nuclear Venture," *Wall Street Journal*, January 1, 2019, https://www.wsj.com/articles/trumps-tech-battle-with-china-roils-bill-gates-nuclear-venture-11546360589.

71. Greene, "Trump's Tech Battle with China Roils Bill Gates Nuclear Venture."

10. PROVISIONING THE POPULACE

1. Jeff Mulhollem, "Double Food Production by 2050? Not So Fast," Futurity, February 27, 2017, https://www.futurity.org/food-production-2050-1368582-2.

2. "Appendix A: How Much Land Is Required for a One-Million-Person City?" Replace Capitalism, http://replacecapitalism.com/appendix-a-how-much-land-is-required-for-a-one-million-person-city.

3. Glen E. Friedman, "Why Vegan?" Burning Flags Press, http://burningflags.com/news/why-vegan.

4. "Meat's Large Water Footprint: Why Raising Livestock and Poultry for Meat Is So Resource-Intensive," Foodbank, https://foodtank.com/news/2013/12/why-meat-eats-resources.

5. Melissa C. Lott, "10 Calories In, 1 Calorie Out: The Energy We Spend on Food," *Scientific American*, August 11, 2011, https://blogs.scientificamerican.com/plugged-in/10-calories-in-1-calorie-out-the-energy-we-spend-on-food.

6. US Environmental Protection Agency, "Resources about Brownfields and Urban Agriculture," EPA, https://www.epa.gov/brownfields/resources-about-brownfields-and-urban-agriculture.

7. Jerry Hayes, "Loss of Honey Bee Populations a Threat to U.S. Agriculture," *Southeast Farm Press*, March 14, 2007, https://www.southeastfarmpress.com/loss-honey-bee-populations-threat-us-agriculture.

8. "USDA Defines Food Deserts," *Nutrition Digest* 38, no. 2, http://americannutritionassociation.org/newsletter/usda-defines-food-deserts.

9. Patrick Caughill, "A Film Made of Graphene Makes Filtering Dirty Water Easier Than Ever," Futurism, February 22, 2018, https://futurism.com/film-graphene-filtering-dirty-water-easier-ever.

10. Jeff Dondero, *The Energy Wise Home* (Lanham, MD: Rowman & Littlefield, 2017).

11. Katie Langin, "Millions of Americans Drink Potentially Unsafe Tap Water: How Does Your County Stack Up?" *Science*, February 12, 2018, https://www.sciencemag.org/news/2018/02/millions-americans-drink-potentially-unsafe-tap-water-how-does-your-county-stack.

12. "Animal Agriculture's Impact on Climate Change," Climate Nexus, https://climatenexus.org/climate-issues/food/animal-agricultures-impact-on-climate-change.

13. Joe Loria, "Animal Agriculture Wastes One-Third of Drinkable Water (and 8 Other Facts for World Water Day)," Mercy for Animals, March 21,

2018, https://mercyforanimals.org/animal-agriculture-wastes-one-third-of-drinkable.

14. Henry Fountain, "Building a $325,000 Burger," *New York Times*, May 12, 2013, https://www.nytimes.com/2013/05/14/science/engineering-the-325000-in-vitro-burger.html; P. K. Thornton, "Livestock Production: Recent Trends, Future Prospects," *Philosophical Transactions B* 365, no. 1554 (September 27, 2010), https://www.ncbi.nlm.nih.gov/pmc/articles/PMC2935116; Mark Tran, "Greenhouse Gas Emissions from Livestock Can Be Cut by 30%, Says FAO," *The Guardian*, September 26, 2013, https://www.theguardian.com/global-development/2013/sep/26/greenhouse-gas-emissions-livestock.

15. Monica Saavoss, "How Might Cellular Agriculture Impact the Livestock, Dairy, and Poultry Industries?" *Choices*, 2019, http://www.choicesmagazine.org/choices-magazine/submitted-articles/how-might-cellular-agriculture-impact-the-livestock-dairy-and-poultry-industries.

16. Fountain, "Building a $325,000 Burger."

17. Kristin Houser, "For the First Time, a Startup Grew a Steak in a Lab," Futurism, December 13, 2018, https://futurism.com/THE-BYTE/LAB-GROWN-STEAK-ALEPH-FARMS.

18. Jenny Splitter, "How Do They Make Meat-Like Burgers from Plants?" Curiosity, May 11, 2018, https://curiosity.com/topics/how-do-they-make-meat-like-burgers-from-plants-curiosity.

19. Houser, "For the First Time, a Startup Grew a Steak in a Lab."

20. Splitter, "How Do They Make Meat-Like Burgers from Plants?"

21. Stefan M. Pasiakos et al., "Sources and Amounts of Animal, Dairy, and Plant Protein Intake of US Adults in 2007–2010," *Nutrients* 7, no. 8 (August 2015): 7058–69, https://www.ncbi.nlm.nih.gov/pmc/articles/PMC4555161.

22. Leo Galland, "The Standard American Diet (SAD)," Leo Galland MD (blog), http://drgalland.com/the-standard-american-diet-sad.

23. K. C. Wright, "The Coup in the Dairy Aisle," *Today's Dietitian* 20, no. 9 (September 2018), https://www.todaysdietitian.com/newarchives/0918p28.shtml.

24. Isabella Grandic, "How to Make Dairy without Cows," Medium, November 3, 2018, https://medium.com/@igrandic03/how-to-make-dairy-without-cows-5bf25bc24dd.

25. Beth Kowitt, "Future of Milk? Genetically Engineered Yeast Could Replace Cows," Genetic Literacy Project, March 21, 2017, https://geneticliteracyproject.org/2017/03/21/future-milk-genetically-engineered-yeast-replace-cows.

26. Samantha Cassetty, "Plant-Based Milk vs. Cow's Milk: What's the Difference?" NBC News, updated August 16, 2018, https://www.nbcnews.com/better/health/plant-based-milk-vs-cow-s-milk-what-s-difference-ncna845271.

27. Tom Ward, "A Team of Scientists Just Made Food from Electricity—and It Could Be the Solution to World Hunger," Futurism, July 26, 2017, https://futurism.com/a-team-of-scientists-just-made-food-from-electricity-and-it-could-be-the-solution-to-world-hunger.

28. "Vertical Farming: Indoor Agriculture," Basic Knowledge 101, http://www.basicknowledge101.com/subjects/verticalfarming.html.

29. Victor Mendez Perez, "Study of the Sustainability Issues of Food Production Using Vertical Farm Methods in an Urban Environment within the State of Indiana" (master's thesis, Purdue University, 2014), https://docs.lib.purdue.edu/dissertations/AAI1565090.

30. Paul Marks, "Vertical Farms Sprouting All over the World," *New Scientist*, January 15, 2014, https://www.newscientist.com/article/mg22129524-100-vertical-farms-sprouting-all-over-the-world.

31. Ronald Holden, "It's Called Vertical Farming, and It Could Be the Future of Agriculture," *Forbes*, November 4, 2017, https://www.forbes.com/sites/ronaldholden/2017/11/04/its-called-vertical-farming-and-it-could-be-the-future-of-agriculture.

32. "Hydroponic Questions," Bolton Farms, http://boltonhydroponics.com/FAQ.php.

33. "Hydroponic Systems—A Way to Save Water," Energy in Water, https://www.energyinwater.eu/hydroponic-systems-a-way-to-save-water.

34. "Hydroponics Yield," Uponics, https://uponics.com/hydroponics-yield.

35. Dom Galeon, "Your Produce Might Soon Grow in a Warehouse Down the Block," Futurism, September 7, 2017, https://futurism.com/your-produce-might-soon-grow-in-a-warehouse-down-the-block.

36. Brian Barth, "How Does Aeroponics Work?" Modern Farmer, July 26, 2018, https://modernfarmer.com/2018/07/HOW-DOES-AEROPONICS-WORK.

37. "What Is Aquaponics?" Aquaponic Source, https://www.theaquaponicsource.com/what-is-aquaponics.

38. Tim Heath and Yiming Shao, "Vertical Farms Offer a Bright Future for Hungry Cities," The Conversation, July 21, 2014, http://theconversation.com/vertical-farms-offer-a-bright-future-for-hungry-cities-26934.

39. Tamara Duker Freuman, "Who Actually Needs a 2,000-Calorie Diet?" *U.S. News*, June 14, 2016, https://health.usnews.com/health-news/blogs/eat-run/articles/2016-06-14/who-actually-needs-a-2-000-calorie-diet.

40. X. Qin et al., "Design of Solar Optical Fiber Lighting System for Enhanced Lighting in Highway Tunnel Threshold Zone: A Case Study of Huashuyan Tunnel in China," *International Journal of Photoenergy* 2015, http://dx.doi.org/10.1155/2015/471364.

41. Captain Paul Watson , "V," Sea Shepherd, https://seashepherd.org/2014/05/06/v.

42. Jean-Michel Cousteau with Jaclyn Mandoske, "The Future of Sustainable Fish Farming," *Ocean Futures Society* (blog), March 17, 2014, http://www.oceanfutures.org/news/blog/future-sustainable-fish-farming.

43. "Mariculture (History of Aquaculture/Lecture 1)," Quizlet, https://quizlet.com/152705306/mariculture-history-of-aquaculturelecture-1-flashcards.

44. Brendan Smith, "The Coming Green Wave: Ocean Farming to Fight Climate Change," *The Atlantic*, November 23, 2011, https://www.theatlantic.

com/international/archive/2011/11/the-coming-green-wave-ocean-farming-to-fight-climate-change/248750.

45. J. P. S. Cabral, "Water Microbiology: Bacterial Pathogens and Water," *International Journal of Environmental Research and Public Health* 7, no. 10 (October 2017): 3657–703, https://www.ncbi.nlm.nih.gov/pmc/articles/PMC2996186.

46. Smith, "The Coming Green Wave."

47. L. A. Helfrich and George Libey, "Fish Farming in Recirculating Aquaculture Systems (RAS)," Texas A&M, September 2013, http://fisheries.tamu.edu/files/2013/09/Fish-Farming-in-Recirculating-Aquaculture-Systems-RAS.pdf.

48. Brendan Smith, "Can Ocean Farms Actually Be More Sustainable Than Even the Most Environmentally Sensitive Traditional Farms?" AlterNet, November 30, 2011, https://www.alternet.org/2011/11/can_ocean_farms_actually_be_more_sustainable_than_even_the_most_environmentally_sensitive_traditional_farms.

49. "Seaweed Aquaculture: An Answer to Sustainable Food and Fuel?" ThinkProgress, December 1, 2011, https://thinkprogress.org/seaweed-aquaculture-an-answer-to-sustainable-food-and-fuel-f60643f701dc.

50. Smith, "The Coming Green Wave."

51. Adrian Oaks, "Spirulina & Chlorella," Fish Oil Facts, January 15, 2016, http://www.fishoilfacts.net/spirulina-chlorella.

52. Smith, "The Coming Green Wave."

53. Hawken, *Drawdown*.

54. Rich McEachran, "Under the Sea: The Underwater Farms Growing Basil, Strawberries and Lettuce," *The Guardian*, August 13, 2015, https://www.theguardian.com/sustainable-business/2015/aug/13/food-growing-underwater-sea-pods-nemos-garden-italy.

55. McEachran, "Under the Sea."

56. K. Moorhead and B. Capelli with G. Cysewski, *Spirulina: Nature's Superfood* (Kailua-Kona, HI: Cyanotech, 1993), https://www.terapiaclark.es/Docs/spirulina_book.pdf.

57. Lauren Cox, "Spirulina: Nutrition Facts & Health Benefits," Live Science, February 6, 2018, https://www.livescience.com/48853-spirulina-supplement-facts.html.

58. Rebecca Rupp, "Make Way for Algae on Your Dinner Plate," *National Geographic*, September 4, 2015, https://www.nationalgeographic.com/people-and-culture/food/the-plate/2015/09/04/make-way-for-algae-on-your-dinner-plate.

59. Hawken, *Drawdown*; Wikipedia, s.v. "Algae Fuel," last modified July 20, 2019, 9:13, https://en.wikipedia.org/wiki/Algae_fuel.

60. Jolene Creighton, "Floating Vertical Farms: Feeding Earth's Growing Population," Futurism, September 10, 2014, https://futurism.com/floating-vertical-farms-feeding-earths-growing-population.

61. Cathy Seigner cites the *Wall Street Journal* in "Startups Are Using Tomatoes, Eggplant and Carrots to Improve Fake Fish," Food Dive, October 18,

2018, https://www.fooddive.com/news/startups-are-using-tomatoes-eggplant-and-carrots-to-improve-fake-fish/539853.

62. Pew Research Center, *Public and Scientists' Views on Science and Society* (Pew Research Center, January 29, 2015), https://www.pewinternet.org/2015/01/29/public-and-scientists-views-on-science-and-society.

63. Kate Siegel and Suzanne Verity, "What You Need to Know about GMOs," WebMD, April 8, 2016, https://www.webmd.com/food-recipes/features/truth-about-gmos#1.

64. Paul McDivitt, "Does GMO Corn Increase Crop Yields? 21 Years of Data Confirm It Does," Genetic Literacy Project, February 19, 2018, https://geneticliteracyproject.org/2018/02/19/gmo-corns-yield-human-health-benefits-vindicated-21-years-studies.

65. "Reports of the Council on Science and Public Health," AMA, 2012, https://www.ama-assn.org/sites/ama-assn.org/files/corp/media-browser/public/hod/a12-csaph-reports_0.pdf.

66. World Health Organization, "Frequently Asked Questions on Genetically Modified Foods," WHO, May 2014, https://www.who.int/foodsafety/areas_work/food-technology/faq-genetically-modified-food/en.

67. Siegel and Verity, "What You Need to Know about GMOs."

68. US Food and Drug Administration, "Center for Veterinary Medicine," FDA, last modified August 29, 2018, https://www.fda.gov/about-fda/office-foods-and-veterinary-medicine/center-veterinary-medicine.

69. Julie Taylor, "10 Ways to Keep Your Diet GMO-Free," CNN. March 31, 2014, https://www.cnn.com/2014/03/25/health/upwave-gmo-free-diet/index.html.

70. American Public Health Association, "Opposition to the Use of Hormone Growth Promoters in Beef and Dairy Cattle Production," APHA, November 10, 2009, https://www.apha.org/policies-and-advocacy/public-health-policy-statements/policy-database/2014/07/09/13/42/opposition-to-the-use-of-hormone-growth-promoters-in-beef-and-dairy-cattle-production.

71. "Non-Hodgkins Lymphoma from Roundup," My Cancer Lawsuit, https://www.mycancerlawsuit.com/non-hodgkins-lymphoma.

72. Jef Feeley, Joel Rosenblatt, and Tim Loh, "Bayer Wants to Settle Roundup Cancer Claims for $8 Billion, Sources Say," *Los Angeles Times*, August 9, 2019, https://www.latimes.com/business/story/2019-08-09/bayer-wants-to-settle-roundup-cancer-claims-for-8-billion-sources-say.

73. Kristin Houser, "US Farmers Can Now Grow Edible Cotton," Futurism, October 18, 2018, https://futurism.com/the-byte/edible-cottonseeds-us-regulations.

74. Wikipedia, s.v. "Applications of 3D Printing," last modified July 24, 2019, 6:24, https://en.wikipedia.org/wiki/Applications_of_3D_printing.

75. "How 3D Food Printing Technology Changing the Way We Eat," *Zazengo* (blog), December 1, 2018, https://www.zazengo.com/3d-food-printing-technology.

76. John Straw, "Why 3D Printed Food Is the Future," *Disruption*, November 24, 2015, https://disruptionhub.com/disrupted-food-why-3d-printed-food-is-the-future-of-food.

77. Jacopo Prisco, "'Foodini' Machine Lets You Print Edible Burgers, Pizza, Chocolate," CNN, updated December 31, 2014, https://www.cnn.com/2014/11/06/tech/innovation/foodini-machine-print-food/index.html.

78. Prisco, "'Foodini' Machine Lets You Print Edible Burgers, Pizza, Chocolate."

79. Laurie Segall, "This 3D Printer Makes Edible Food," CNN Money, January 24, 2011, https://money.cnn.com/2011/01/24/technology/3D_food_printer/index.htm.

80. Ken Tudor, "The Case for Insect Protein in Foods," *Daily Vet* (blog), PetMD, January 23, 2014, https://www.petmd.com/blogs/thedailyvet/kentudor/2014/january/insects-pet-food-protein-future-31265.

81. Joseph Bennington-Castro, "How Crickets Could Help Save the Planet," NBC News, February 16, 2017, https://www.nbcnews.com/mach/environment/how-eating-crickets-could-help-save-planet-n721416.

82. Kai Kupferschmidt, "Why Insects Could Be the Ideal Animal Feed," *Science*, October 14, 2015, https://www.sciencemag.org/news/2015/10/feature-why-insects-could-be-ideal-animal-feed.

83. Mid-Atlantic Fishery Management Council, *Atlantic Mackerel, Squid, and Butterfish Fisheries Fisheries Management Plan (FMP), Amendment No. 5, Exclusive Economic Zone (EEZ) US Atlantic Coast: Environmental Impact Statement* (1995), https://books.google.com/books?id=3j43AQAAMAAJ.

84. "Why You Should Eat Insects: Cricket vs. Beef," *Näakbar* (blog), https://naakbar.com/blogs/articles/why-you-should-eat-insect-cricket-versus-beef.

85. Ken Tudor, " Insect Protein: Is It a Viable Alternative for Pet and Livestock Food?" Natural Products Insider, March 3, 2015, https://www.naturalproductsinsider.com/ingredients/insect-protein-it-viable-alternative-pet-and-livestock-food.

86. "Will We All Be Eating Insects in 50 Years?" IFLScience! https://www.iflscience.com/environment/will-we-all-be-eating-insects-50-years.

87. Centers for Disease Control and Prevention, "Zoonotic Diseases," CDC, https://www.cdc.gov/onehealth/basics/zoonotic-diseases.html.

11. SMART CITY SYSTEMS

1. "Smart Home or Building (Home Automation or Domotics)," IoT Agenda, last modified July 2018, https://internetofthingsagenda.techtarget.com/definition/smart-home-or-building.

2. "You're Using Your Fridge Wrong," CNET, https://www.cnet.com/pictures/youre-using-your-refrigerator-wrong-pictures.

3. "D6T MEMS Thermal Sensors," OMRON, https://www.components.omron.com/product-detail?partNumber=D6T.

4. Sam Biddle, "For Owners of Amazon's Ring Security Camera, Strangers May Have Been Watching Too," The Intercept, January 10, 2019, https://theintercept.com/2019/01/10/amazon-ring-security-camera.

5. Wikipedia, s.v. "X10 (Industry Standard)," last modified July 14, 2019, 13:30, https://en.wikipedia.org/wiki/X10_(industry_standard).

6. Prachi Bhardwaj, "Disney's 'Smart House' Came Out in 1999—Here's All the Technology from the Movie That We're Actually Using 2 Decades Later," *Business Insider*, April 26, 2018. https://www.businessinsider.com/disney-smart-house-tech-smart-home-technologies-2018-4.

7. Drew Hendricks, "The History of Smart Homes," IoT Evolution, April 22, 2014, https://www.iotevolutionworld.com/m2m/articles/376816-history-smart-homes.htm.

8. Molly Edmunds and Nathan Chandler, "How Smart Homes Work," HowStuffWorks, March 25, 2008, https://home.howstuffworks.com/smart-home.htm.

9. Kinsey Grant, "How to Build Your Home Just Like Microsoft Billionaire Bill Gates," The Street, updated June 28, 2018, https://www.thestreet.com/story/14321844/1/how-to-build-your-home-just-like-billionaire-bill-gates.html.

10. "Internet of Things (IoT)," IoT Agenda, last modified July 2019, https://internetofthingsagenda.techtarget.com/definition/Internet-of-Things-IoT.

11. Jeff Dondero, *The Energy Wise Workplace* (Lanham, MD: Rowman & Littlefield, 2017).

12. Barry Brook, "Energy Storage Discussion Thread," *Brave New Climate* (blog), November 13, 2011, https://bravenewclimate.com/2011/11/13/energy-storage-dt.

13. Dondero, *Energy Wise Workplace*.

14. "Internet of Things (IoT)."

15. Renée Lynn Midrack, "What Is a Smart Lock and Why Would You Want One?" Lifewire, last modified June 5, 2019, https://www.lifewire.com/smart-locks-4159894.

16. "How Much Does It Cost to Repair & Cleanup Water Damage?" Home Advisor, https://www.homeadvisor.com/cost/disaster-recovery/repair-water-damage.

17. Martin, "The Smart Home: Intelligent Home Automation," Cleverism, August 22, 2014, https://www.cleverism.com/smart-home-intelligent-home-automation; Eliot, "9 Ways a Smart Home Can Improve Your Life," Smart-Things (blog), March 31, 2015, https://blog.smartthings.com/news/roundups/9-ways-a-smart-home-can-improve-your-life.

18. Todd Jaquith, "Here's a Look at the Smart Cities of the Future," Futurism, January 18, 2017, https://futurism.com/heres-a-look-at-the-smart-cities-of-the-future.

19. Jaquith, "Here's a Look at the Smart Cities of the Future."

20. Alvaro Sanchez-Miralles et al., "Use of Renewable Energy Systems in Smart Cities," in *Use, Operation and Maintenance of Renewable Energy Systems in Smart Cities*, ed. Miguel A. Sanz-Bobi (Cham, Switzerland: Springer), 341–70, https://link.springer.com/chapter/10.1007/978-3-319-03224-5_10.

21. P. Sethi and Smruti R. Sarangi, "Internet of Things: Architectures, Protocols, and Applications," *Journal of Electrical and Computer Engineering* 2017, https://www.hindawi.com/journals/jece/2017/9324035; Sekhar Kondepudi, "An Overview of Smart Sustainable Cities and the Role of Information and Communication Technologies (ICTs)," ITU, May 3, 2015, https://www.itu.int/en/ITU-D/Regional-Presence/ArabStates/Documents/events/2015/SSC/S3-DrSekharKondepudi.pdf.

22. Sy Mukherjee, "Prepare for the Digital Health Revolution," *Fortune*, April 20, 2017, https://fortune.com/2017/04/20/digital-health-revolution; World Economic Forum, *Health Systems Leapfrogging in Emerging Economies: Project Paper*, WEForum, January 2014, http://www3.weforum.org/docs/WEF_HealthSystem_LeapfroggingEmergingEconomies_ProjectPaper_2014.pdf.

23. National Center for Chronic Disease Prevention and Health Promotion, Division for Heart Disease and Stroke Prevention, "Atrial Fibrillation Fact Sheet," CDC, https://www.cdc.gov/dhdsp/data_statistics/fact_sheets/fs_atrial_fibrillation.htm.

24. Mirza Baig and Hamid Gholamhosseini, "Smart Health Monitoring Systems: An Overview of Design and Modeling," *Journal of Medical Systems* 37, no. 2 (April 2013), https://www.researchgate.net/publication/234142160_Smart_Health.

25. Ellie Zolfagharifard, "Smart Toilet That Can Analyse Your PEE and an App That Detects Depression: Japanese Expo Reveals Strange Health Gadgets," *Daily Mail Online*, November 5, 2015, https://www.dailymail.co.uk/sciencetech/article-3306193/Smart-toilet-analyse-PEE-app-detects-depression-Japanese-expo-reveals-strange-health-gadgets.html.

26. Christian de Looper, "Google's Smart Bathroom Patent Puts Sensors in Your Toilet, Tub, and Mirror," Digital Trends, August 10, 2016, https://www.digitaltrends.com/home/google-smart-bathroom-patent; Stacy Liberatore, "Smart Toilets That Analyse Your Poop and Ultrasonic Baths That 3D Scan Your Internal Organs: Google Patent Reveals Plan for Bathroom That Can Monitor Your Health," *Daily Mail Online*, August 1, 2016, https://www.dailymail.co.uk/sciencetech/article-3718788/Smart-toilets-analyse-poop-ultrasonic-baths-3D-scan-internal-organs-Google-patent-reveals-plan-bathroom-monitor-health.html.

27. James Heskett, "Does Internet Technology Threaten Brand Loyalty?" Working Knowledge, June 2, 2014, https://hbswk.hbs.edu/item/does-internet-technology-threaten-brand-loyalty.

28. "How Much Does It Cost to Install a Home Automation System?" HomeAdvisor, https://www.homeadvisor.com/cost/electrical/install-or-repair-a-home-automation-system.

29. Luke Denne, Greg Sadler, and Makda Ghebreslassie, "We Hired Ethical Hackers to Hack a Family's Smart Home: Here's How It Turned Out," CBC, September 30, 2018, https://www.cbc.ca/news/technology/smart-home-hack-marketplace-1.4837963.

30. Danny Thakkar, "Problems with Current Security Systems That You Should Know," Bayometric, https://www.bayometric.com/problems-current-security-systems; *Forbes* Technology Council, "13 Factors to Consider with Smart Home Products," *Forbes*, January 23, 2018, https://www.forbes.com/sites/forbestechcouncil/2018/01/23/13-factors-to-consider-with-smart-home-products.

31. Martin, "The Smart Home."

32. Alex Young, "What Is a Security System and How Does it Work?" Safewise, last modified May 9, 2019, https://www.safewise.com/home-security-faq/how-do-security-systems-work.

33. Brian Ross, "Boston Bombing Day 2: The Improbable Story of How Authorities Found the Bombers in the Crowd," ABC News, April 19, 2016, https://abcnews.go.com/US/boston-bombing-day-improbable-story-authorities-found-bombers/story?id=38375726.

34. Caroline Mortimer, "London Council at Centre of CCTV Row Claims Presence of Cameras Could Attract Terrorists Seeking Publicity," *The Independent*, June 4, 2016, https://www.independent.co.uk/news/uk/crime/london-council-at-centre-of-cctv-row-claims-cameras-could-encourage-publicity-seeking-terrorists-a7064486.html.

35. Elina Shatkin, "What a Country! Yakov Smirnoff Is Back . . . and He's a Love Guru," *Los Angeles*, September 27, 2013, https://www.lamag.com/culturefiles/what-a-country-yakov-smirnoff-is-back-and-hes-a-love-guru.

36. Akshay Ghone, "ERP Software Market Is Expected to Reach $41.69 Billion, Globally, by 2020," Allied Market Research, March 2015, https://www.alliedmarketresearch.com/press-release/global-ERP-software-market-is-expected-to-reach-41-69-billion-by-2020.html.

37. Zion Market Research, "Global Smart Home Market to Exceed $53.45 Billion by 2022," GlobalNewswire, January 3, 2018, https://www.globenewswire.com/news-release/2018/01/03/1281338/0/en/Global-Smart-Home-Market-to-Exceed-53-45-Billion-by-2022-Zion-Market-Research.html.

38. "IoT Connections to Grow 140% to Hit 50 Billion by 2022, as Edge Computing Accelerates ROI," Juniper Research, https://www.juniperresearch.com/press/press-releases/iot-connections-to-grow-140-to-hit-50-billion.

39. Ipshita Biswas, "Smart Homes: Past, Present, and Future," Colocation America, December 18, 2018, https://www.colocationamerica.com/blog/smart-homes-past-present-future.

40. Ron Miller, "Cheaper Sensors Will Fuel the Age of Smart Everything," Tech Crunch, March 10, 2015, https://techcrunch.com/2015/03/10/cheaper-sensors-will-fuel-the-age-of-smart-everything; Michael Caccavale, "The Impact of the Digital Revolution on the Smart Home Industry," *Forbes*, September 24, 2018, https://www.forbes.com/sites/forbesagencycouncil/2018/09/24/the-impact-of-the-digital-revolution-on-the-smart-home-industry.

41. "Home Automation Cost," Fixr, https://www.fixr.com/costs/home-automation.

42. Sveta McShane and Jason Dorrier, "Ray Kurzweil Predicts Three Technologies Will Define Our Future," SingularityHub, April 19, 2016, https://

singularityhub.com/2016/04/19/ray-kurzweil-predicts-three-technologies-will-define-our-future.

43. Christianna Reedy, "When Will Humanoid Robots Enter Our Homes and Transform Our Lives," Futurism, June 1, 2017, https://futurism.com/when-will-humanoid-robots-enter-our-homes-and-transform-our-lives; Aaron Smith and Janna Anderson, "Predictions for the State of AI and Robotics in 2025," Pew Research Center, August 6, 2014, https://www.pewinternet.org/2014/08/06/predictions-for-the-state-of-ai-and-robotics-in-2025.

44. Francis X. Shen, "Sex Robots Are Here, but Laws Aren't Keeping Up with the Ethical and Privacy Issues They Raise," The Conversation, February 12, 2019, https://theconversation.com/sex-robots-are-here-but-laws-arent-keeping-up-with-the-ethical-and-privacy-issues-they-raise-109852.

45. Rhian Morgan, "Looking for Robot Love? Here Are 5 Sexbots You Can Buy Right Now," Metro, September 13, 2017, https://metro.co.uk/2017/09/13/looking-for-robot-love-here-are-5-sexbots-you-can-buy-right-now-6891378.

46. James McPhail, "Bottom Line and Beyond: Smarter Commercial Energy Management," Forbes, https://www.forbes.com/sites/forbestechcouncil/2017/11/07/bottom-line-and-beyond-smarter-commercial-energy-management.

47. "The Pros and Cons of Home Automation Systems," ThinkEnergy, https://www.mythinkenergy.com/pros-cons-home-automation; "What Are the Pros and Cons of Different Smart Home Systems?" Vivint, August 3, 2018, https://www.vivint.com/resources/article/pros-and-cons-of-smart-home-systems; Alexis Writing, "Pros & Cons of Smart Home Technology," Hunker, https://www.hunker.com/13401102/pros-cons-of-smart-home-technology.

48. Jeff St. John, "The Networked Grid: Smart Grid, Meet Smart Buildings," Greentech Media, March 1, 2012, https://www.greentechmedia.com/articles/read/the-networked-grid-smart-grid-meet-smart-buildings; US Department of Energy, "Smart Home and Building Systems," NREL, https://www.nrel.gov/esif/smart-home-building-systems.html; Yoeba Penya et al., "Smart Buildings and the Smart Grid," ResearchGate, November 2013, https://www.researchgate.net/publication/258241122_Smart_Buildings_and_the_Smart_Grid.

12. SUPER SKYSCRAPERS

1. Department of Economic and Social Affairs of the United Nations Secretariat, World Economic and Social Survey 2013: Sustainable Development (New York: United Nations, 2013), https://sustainabledevelopment.un.org/content/documents/2843WESS2013.pdf; Pascale Céron and Pascale Gorges-Levard, eds., Sustainable Land-Use Planning and Construction, trans. Katherine Parks (Pantin, France: ARENE Île-de-France, 2015), https://www.areneidf.org/sites/default/files/arene_44p_english_bat_bas_def1.pdf; Sekhar N. Kondepudi et al., Smart Sustainable Cities: An Analysis of Definitions (ITU,

2014), https://www.itu.int/en/ITU-T/focusgroups/ssc/Documents/Approved_Deliverables/TR-Definitions.docx.

2. Shirley Li, "The Skyscraper of the Future," *The Atlantic*, May 18, 2015, https://www.theatlantic.com/technology/archive/2015/05/the-skyscraper-of-the-future/387118.

3. "Vertical Cities Project Benefits," MIPL, http://www.mipl.ind.in/about-us.html.

4. Wikipedia, s.v. "Height of Buildings Act of 1910," last modified December 8, 2018, 22:20, https://en.wikipedia.org/wiki/Height_of_Buildings_Act_of_1910.

5. Wikipedia, s.v. "Height of Buildings Act of 1910."

6. Wikipedia, s.v. "Early Skyscrapers," last modified July 21, 2019, 3:53, https://en.wikipedia.org/wiki/Early_skyscrapers.

7. Blair Kamin, "Frank Lloyd Wright's Mile-High Skyscraper Never Built, but Never Forgotten," *Chicago Tribune*, May 28, 2017, https://www.chicagotribune.com/columns/ct-frank-lloyd-wright-mile-high-met-0528-20170528-column.html.

8. Kelsey Campbell-Dollaghan, "Can China Really Build the World's Tallest Building in 90 Days?" Gizmodo, May 30, 2013, https://gizmodo.com/chinas-radical-plan-to-build-the-worlds-tallest-build-510487766.

9. Stephy Chung, "Dizzying Heights: Tokyo's Future Skyline Could Include a Mile-High Skyscraper," CNN, last modified November 16, 2016, https://www.cnn.com/style/article/tokyo-mile-high-skyscraper/index.html.

10. Li, "The Skyscraper of the Future."

11. Nate Berg, "Is There a Limit to How Tall Buildings Can Get?" CityLab, August 16, 2012, https://www.citylab.com/design/2012/08/there-limit-how-tall-buildings-can-get/2963.

12. Berg, "Is There a Limit to How Tall Buildings Can Get?"

13. Berg, "Is There a Limit to How Tall Buildings Can Get?"

14. Rachel Ross, "Ups & Downs: The Evolution of Elevators," Live Science, December 21, 2016, https://www.livescience.com/57282-elevator-history.html.

15. Edward Glaeser, *Triumph of the City: How Our Greatest Invention Makes Us Richer, Smarter, Greener, Healthier, and Happier* (New York: Penguin Books, 2012).

16. "Elevator," Great Idea Finder, last modified March 26, 2007, http://www.ideafinder.com/history/inventions/elevator.htm.

17. William J. Angelo, John T. Harding, and Andy Kunz, "Rail Race: Maglev or HSR, What Is the Future? Join the Rail Debate and Help Shape History," *Engineering News-Record*, July 21, 2009, https://www.enr.com/articles/8193-rail-race-maglev-or-hsr-what-is-the-future-join-the-rail-debate-and-help-shape-history?v=preview.

18. Emily Nonko, "Can This Elevator Help Designers Sidestep Tall-Building Problems?" CityLab, July 12, 2017, https://www.citylab.com/design/2017/07/elevator-of-the-future-travels-sideways/533316.

19. Nick Paumgarten, "Up and Then Down," *New Yorker*, April 21, 2008, https://www.newyorker.com/magazine/2008/04/21/up-and-then-down.

20. Jenni Marsh and Jane Sit, "Shanghai Tower Picks Up 3 Guinness World Records Including Fastest Elevator," CNN, last modified April 19, 2017, https://www.cnn.com/style/article/worlds-fastest-tower/index.html.

21. Associated Press, "Super Tall, Super Skinny New York Buildings Grow," *Asbury Park Press*, March 6, 2016 , https://www.app.com/story/money/business/2016/03/06/super-tall-super-skinny-new-york-buildings-grow/80865748.

22. Li, "The Skyscraper of the Future."

23. Natalie J. Park, "Tall Building Delirium: The Second Life of the Metlife Building" (PhD diss., University of Hawai'i at Mānoa, 2016), https://scholarspace.manoa.hawaii.edu/bitstream/10125/45576/Park_Natalie_Spring%202016.pdf.

24. Matthew Crosby, "Will There Ever Be an Airbnb or Uber for the Electricity Grid?" Greentech Media, September 8, 2014, https://www.greentechmedia.com/articles/read/an-airbnb-or-uber-for-the-electricity-grid#gs.tkn5if; "Renewable Energy and Electricity," World Nuclear Association, last modified May 2019, https://www.world-nuclear.org/information-library/energy-and-the-environment/renewable-energy-and-electricity.aspx.

25. "Innovation 2050: A Digital Future for the Infrastructure Industry," Balfour Beatty, https://www.balfourbeatty.com/2050.

26. Luke Tsai, "Architect Eugene Tssui Might Be the Most Interesting Man in the East Bay," *East Bay Express*, January 31, 2017, https://www.eastbayexpress.com/oakland/architect-eugene-tssui-might-be-the-most-interesting-man-in-the-east-bay/Content?oid=5098108.

27. Wikipedia, s.v. "Space Elevator," last modified August 4, 2019, 2:06, https://en.wikipedia.org/wiki/Space_elevator; Tom Nardi, "One Small Step for a Space Elevator," Hackaday, September 4, 2018, https://hackaday.com/tag/space-elevator.

28. Gregg D. Ander, "Daylighting," WBDG, last modified September 15, 2016, https://www.wbdg.org/resources/daylighting.

29. "What Is Aquaponics?" Aquaponic Source, https://www.theaquaponicsource.com/what-is-aquaponic.

30. Lulu Chang, "Wish Your Wallpaper Would Evolve with Your Taste? With Lumentile, It Can," Digital Trends, December 19, 2016, https://www.digitaltrends.com/cool-tech/lumentile-digital-wallpaper.

31. Alan Martin, "Forget Wallpaper, You Can Now Turn Your Whole Wall into a TV with Samsung's 146-In. Giant Modular Set," January 8, 2018, https://www.alphr.com/samsung/1008120/samsung-the-wall-146in-tv.

32. "Future of Smart Cities Lies with Communication Technologies," *Infobip* (blog), https://www.infobip.com/en/blog/powering-smart-cities-with-communication-technologies.

33. Aria Bendix, "The Visionary Mega-Tower That San Francisco Never Built," CityLab, October 28, 2015, https://www.citylab.com/design/2015/10/the-visionary-mega-tower-that-san-francisco-never-built/412135.

34. "Shaping the Future of Drone Delivery," Airbus, February 7, 2018, https://www.airbus.com/newsroom/news/en/2018/02/shaping-the-future-of-drone-delivery.html.

35. "Will Maglev Ever Become Mainstream?" Railway Technology, January 17, 2018, https://www.railway-technology.com/features/will-maglev-ever-become-mainstream.

36. Alex Gray, "Countries Are Announcing Plans to Phase Out Petrol and Diesel Cars. Is Yours on the List?" World Economic Forum, September 26, 2017, https://www.weforum.org/agenda/2017/09/countries-are-announcing-plans-to-phase-out-petrol-and-diesel-cars-is-yours-on-the-list.

37. Timothy Bralower and David Bice, "Heat Capacity and Energy Storage," Earth 103: Earth in the Future, https://www.e-education.psu.edu/earth103/node/1005.

38. Christian Müller, "Cities of the Future," AXA XL, June 2010, https://xlgroup.com/~/media/bf5909d23e244150bdacd203fbf69ca5.pdf.

39. "Tuned Mass Damper of Taipei 101," Atlas Obscura, https://www.atlasobscura.com/places/tuned-mass-damper-of-taipei-101.

13. NEW USE FOR REFUSE

1. Ann Simmons, "The World's Trash Crisis, and Why Many Americans Are Oblivious," *Los Angeles Times*, April 22, 2016, https://www.latimes.com/world/global-development/la-fg-global-trash-20160422-20160421-snap-htmlstory.html.

2. Oliver Balch., "The Future of Waste: Five Things to Look For by 2025," *The Guardian*, February 23, 2015, https://www.theguardian.com/sustainable-business/2015/feb/23/future-of-waste-five-things-look-2025.

3. Rebecca Smithers, "Almost Half of the World's Food Thrown Away, Report Finds," *The Guardian*, January 10, 2013, https://www.theguardian.com/environment/2013/jan/10/half-world-food-waste.

4. US Environmental Protection Agency, "Sustainable Management of Construction and Demolition Materials," EPA, https://www.epa.gov/smm/sustainable-management-construction-and-demolition-materials.

5. Myles Gough cites the *International Business Times* in "World's Tallest 3D-Printed Building Showcased in China," ScienceAlert, January 29, 2015, https://www.sciencealert.com/world-s-tallest-3d-printed-building-showcased-in-china.

6. Katharine Schwab, "The Building Materials of the Future Are . . . Old Buildings," *Fast Company*, May 5, 2018, https://www.fastcompany.com/90159252/the-building-materials-of-the-future-are-old-buildings.

7. Chelsea Harvey, "This Is How Cities of the Future Will Get Their Energy," *Washington Post*, May 20, 2016, https://www.washingtonpost.com/news/energy-environment/wp/2016/05.

8. United Nations Department of Economic and Social Affairs, "Around 2.5 Billion More People Will Be Living in Cities by 2050, Projects New UN

Report," UN, May 16, 2018, https://www.un.org/development/desa/en/news/population/2018-world-urbanization-prospects.html.

9. Patrick Conners, "Methane from Landfills," Energy Forums, March 27, 2016, http://energyforums.net/alternative-energy/methane-from-landfills.

10. Stephen Lacey, "Look at How Much Waste America Puts into Landfills," Greentech Media, June 3, 2013, https://www.greentechmedia.com/articles/read/look-at-how-much-waste-america-puts-into-landfills-compared-to-europe#gs.tn6udy.

11. US Environmental Protection Agency, "Sources of Greenhouse Gas Emissions," EPA, https://www.epa.gov/ghgemissions/sources-greenhouse-gas-emissions.

12. "Strategic Metals: Will Future Supply Be Able to Meet Future Demand?" Mission 2016: The Future of Strategic Natural Resources, http://web.mit.edu/12.000/www/m2016/finalwebsite/solutions/landfill.html.

13. Michael Blanding, "Transforming Manufacturing Waste into Profit," Working Knowledge, October 3, 2011, https://hbswk.hbs.edu/item/transforming-manufacturing-waste-into-profit.

14. US Environmental Protection Agency, "Mining Waste," EPA, last modified April 19, 2016, https://archive.epa.gov/epawaste/nonhaz/industrial/special/web/html/index-5.html.

15. Jeff Nelson, "How Much Water to Make a Pound of Beef?" VegSource, March 1, 2001, https://www.vegsource.com/articles/pimentel_water.htm.

16. "Consumption and Population: Is California Big Enough?" World Population Awareness, http://www.population-awareness.net/CalifPop.html.

17. Gary Kardys, "High Performance Slag Materials: A Steel Industry By-product," Engineering 360, January 25, 2018, https://insights.globalspec.com/article/7809/high-performance-slag-materials-a-steel-industry-byproduct.

18. US Environmental Protection Agency, "Basic Information about Electronics Stewardship," EPA, last modified October 16, 2018, https://www.epa.gov/smm-electronics/basic-information-about-electronics-stewardship.

19. "Electronic Waste Reaches Record High of 45 Million Tons," DW, December 13, 2017, https://www.dw.com/en/electronic-waste-reaches-record-high-of-45-million-tons/a-41784177.

20. Stephen Leahy, "China's Booming Middle Class Drives Asia's Toxic E-Waste Mountains," *The Guardian*, January 16, 2017, https://www.theguardian.com/environment/2017/jan/16/chinas-booming-middle-class-drives-asias-toxic-e-waste-mountains.

21. Emma Woollacott, "E-Waste Mining Could Be Big Business—and Good for the Planet," BBC News, July 3, 2018, https://www.bbc.com/news/business-44642176.

22. Susanna Kim, "What Happens to Your Recycled iPhones and Other Apple Products," ABC News, March 25, 2016, https://abcnews.go.com/Business/recycled-iphones-apple-products/story?id=37872881.

23. Woollacott, "E-Waste Mining Could Be Big Business."

24. "How Many Precious Metals Are Found in Electronic Devices," *ERI* (blog), June 23, 2015, https://eridirect.com/blog/2015/06/how-many-precious-metals-are-found-in-electronic-devices.

25. "A Brief Guide to Precious Metals in Electronics," Apmex, https://www.apmex.com/education/science/a-brief-guide-to-precious-metals-in-electronics.

26. Transparency Market Research, "Recycled Metal Market Is Estimated to Be Worth US\$476.2 Billion by 2024," Cision, July 11, 2016, https://www.prnewswire.com/news-releases/recycled-metal-market-is-estimated-to-be-worth-us4762-billion-by-2024-global-industry-analysis-size-share-growth-trends-and-forecast-2016---2024-tmr-586315671.html.

27. Matt Kennedy, "Cell Phone Recycling," WasteCare Corporation, https://www.wastecare.com/Articles/Cell_Phone_Recycling.htm.

28. Dolores Hidalgo et al., "Sustainable Vacuum Waste Collection Systems in Areas of Difficult Access," *Tunnelling and Underground Space Technology* 81 (November 2018): 221–27, https://doi.org/10.1016/j.tust.2018.07.026.

29. Jessica Lyons Hardcastle, "Commercial Waste Collection Zones Reduces Truck Traffic, Greenhouse Gas Emissions," Environmental Leader, August 18, 2016, https://www.environmentalleader.com/2016/08/commercial-waste-collection-reduces-truck-traffic-greenhouse-gas-emissions.

30. N. M. Yusof, A. Z. Jidin, and M. I. Rahim, "Smart Garbage Monitoring System for Waste Management," ResearchGate, January 2017, https://www.researchgate.net/publication/313252675_Smart_Garbage_Monitoring_System_for_Waste_Management.

31. M. A. Hannan et al., "Radio Frequency Identification (RFID) and Communication Technologies for Solid Waste Bin and Truck Monitoring System," *Waste Management* 31, no. 12 (December 2011): 2406–13, https://doi.org/10.1016/j.wasman.2011.07.022.

32. Rohan Chaudhari, "Waste Management (Future Cities)." We Sustainable, May 24, 2015, https://wesustainable.wordpress.com/2015/05/24/waste-management-future-cities.

33. Josh Loeb, "Genetically Engineered Slugs to Chew through Landfill and Mine Precious Metals," *E&T*, November 8, 2017, https://eandt.theiet.org/content/articles/2017/11/genetically-engineered-slugs-to-chew-through-landfill-and-mine-precious-metals.

34. Damian Carrington, "Scientists Accidentally Create Mutant Enzyme That Eats Plastic Bottles," *The Guardian*, April 16, 2018, https://www.theguardian.com/environment/2018/apr/16/scientists-accidentally-create-mutant-enzyme-that-eats-plastic-bottles.

35. Loeb, "Genetically Engineered Slugs to Chew through Landfill and Mine Precious Metals."

36. Megan Gannon, "Solid Gold: Poop Could Yield Precious Metals," Live Science, March 24, 2015, https://www.livescience.com/50235-solid-gold-poop-could-yield-precious-metals.html.

37. Sarah Zhang, "There's Millions of Dollars Worth of Gold and Silver in Sewage," Gizmodo, January 16, 2015, https://gizmodo.com/theres-millions-of-dollars-worth-of-gold-and-silver-in-1680046919.

38. Jeff Dondero, *Throwaway Nation: The Ugly Truth about American Garbage* (Lanham, MD: Rowman & Littlefield, 2019); Warren Cornwall, "Sewage Sludge Could Contain Millions of Dollars Worth of Gold," *Science*, January 16, 2015, https://www.sciencemag.org/news/2015/01/sewage-sludge-could-contain-millions-dollars-worth-gold.

39. "Mining for Metals in Society's Waste," IFLScience! https://www.iflscience.com/environment/mining-metals-society-s-waste.

40. Ashley Singh, "Road Lengths in Great Britain 2017," Department for Transport, July 5, 2018, https://assets.publishing.service.gov.uk/government/uploads/system/uploads/attachment_data/file/722478/road-lengths-in-great-britain-2017.pdf.

41. "Frequently Asked Questions," ARTBA, https://www.artba.org/about/faq.

42. "How Recycling Is Changing in All 50 States," Waste Dive, https://www.wastedive.com/news/what-chinese-import-policies-mean-for-all-50-states/510751.

43. "The Future of Robotics in Waste Sorting," *GK Blog*, General Kinematics, https://www.generalkinematics.com/blog/future-robotics-waste-sorting.

44. "What Can Be Recycled? A List of 200+ Items," *Blog*, Personal Creations, https://www.personalcreations.com/blog/how-to-recycle-anything.

45. XiaoZhi Lim, "Turning Organic Waste into Hydrogen," *Chemical Engineering News*, April 7, 2019, https://cen.acs.org/energy/hydrogen-power/Turning-organic-waste-hydrogen/97/i14.

46. "Advanced Space Transportation Program: Paving the Highway to Space," NASA, last modified April 12, 2008, https://www.nasa.gov/centers/marshall/news/background/facts/astp.html.

47. Derek Thompson, "2.6 Trillion Pounds of Garbage: Where Does the World's Trash Go?" *The Atlantic*, June 7, 2012, https://www.theatlantic.com/business/archive/2012/06/26-trillion-pounds-of-garbage-where-does-the-worlds-trash-go/258234.

48. Ethan Siegel, "Ask Ethan: Why Don't We Shoot Earth's Garbage into the Sun?" *Forbes*, October 1, 2016, https://www.forbes.com/sites/startswithabang/2016/10/01/ask-ethan-why-dont-we-shoot-earths-garbage-into-the-sun.

49. Mary Catherine O'Connor, "Only 14% of Plastics Are Recycled: Can Tech Innovation Tackle the Rest?" *The Guardian*, February 22, 2017, https://www.theguardian.com/sustainable-business/2017/feb/22/plastics-recycling-trash-chemicals-styrofoam-packaging.

50. Wikipedia, s.v. "Carton," last modified July 2, 2019, 14:02, https://en.wikipedia.org/wiki/Carton; Eva Portgrácz, "The Environmental Impacts of Packaging," in *Environmentally Conscious Materials and Chemicals Processing*, ed. Myer Kutz (Hoboken, NJ: Wiley, 2007), 237–78, https://www.researchgate.net/publication/229796182_The_Environmental_Impacts_of_Packaging; "A to Z Recycling Guide to Almost Anything and Everything," Green University, https://www.greenuniversity.com/Recycle_Anything_Guide.pdf.

51. "Are Drink Pouches Recyclable or Trash?" *WasteAway* (blog), Waste-Away Group, March 14, 2018, http://wasteawaygroup.blogspot.com/2018/03/are-drink-pouches-recyclable-or-trash.html.

52. Arlene Karidis, "What Will the Future Landfill Look Like?" Waste 360, July 5, 2018, https://www.waste360.com/landfill-operations/what-will-future-landfill-look.

53. Balch, "The Future of Waste."

54. Mark Harris, "Carbon Fibre: The Wonder Material with a Dirty Secret," *The Guardian*, March 22, 2017, https://www.theguardian.com/sustainable-business/2017/mar/22/carbon-fibre-wonder-material-dirty-secret; Peter Suciu, "The Perplexing Carbon Fiber Repurposing Problem," Tech News World, September 18, 2012, https://www.technewsworld.com/story/76172.html.

55. Chang May Choon, "South Korea Cuts Food Waste with 'Pay as You Trash,'" *Straits Times*, April 24, 2016, https://www.straitstimes.com/asia/east-asia/south-korea-cuts-food-waste-with-pay-as-you-trash.

56. Jessica Lyons Hardcastle, "Trash Talk: Is Waste-to-Energy the Next Step in Sustainable Waste Management?" Environmental Leader, April 26, 2016, https://www.environmentalleader.com/2016/04/trash-talk-is-waste-to-energy-the-next-step-in-sustainable-waste-management.

57. Eric Mack, "It's Plastic. It's Edible. It Could Be a Very Big Deal," *Forbes*, September 21, 2018, https://www.forbes.com/sites/ericmack/2018/09/21/algotek-algae-plastic-to-change-the-world-by-disappearing.

58. Bruce V. Bigelow, "San Diego Startup Uses Algae Feedstock to Make Renewable Flip-Flops," Xconomy, October 6, 2017, https://xconomy.com/san-diego/2017/10/06/san-diego-startup-uses-algae-feedstock-to-make-renewable-flip-flops.

59. Wikipedia, s.v. "Anaerobic Digestion," last modified August 1, 2019, 9:08, https://en.wikipedia.org/wiki/Anaerobic_digestion.

60. "What a Waste: An Updated Look into the Future of Solid Waste Management," World Bank, September 20, 2018, https://www.worldbank.org/en/news/immersive-story/2018/09/20/what-a-waste-an-updated-look-into-the-future-of-solid-waste-management.

14. CAUTIONARY COMMENTS ON CLIMATE CHANGE

1. M. Marchand, TrinhThi Long, and Sawarendro, *Adaptive Water Management for Delta Regions* (Deltares, 2012), http://publications.deltares.nl/1205471_000.pdf.

2. 4 Ways Permeable Paving Provides Better Stormwater Management. *Truegrid* (blog), https://www.truegridpaver.com/blog/4-ways-permeable-paving-provides-better-stormwater-management.

3. Tim Folger, "Rising Seas," *National Geographic*, September 2013, https://www.nationalgeographic.com/magazine/2013/09/rising-seas-coastal-impact-climate-change; Jen Schwartz, "Surrendering to Rising Seas," *Scientific*

American, August 1, 2018, https://www.scientificamerican.com/article/
surrendering-to-rising-seas.

4. Union of Concerned Scientists, "Published Study Identifies When Hundreds of Coastal Communities Will Face Inundation, Possible Retreat," UCSUSA, July 12, 2017. https://www.ucsusa.org/press/2017/study-identifies-when-coastal-communities-will-face-inundation-possible-retreat.

5. Joseph Bennington-Castro, "Walls Won't Save Our Cities from Rising Seas: Here's What Will," NBC News, July 27, 2017, https://www.nbcnews.com/mach/science/walls-won-t-save-our-cities-rising-seas-here-s-ncna786811.

6. Bennington-Castro, "Walls Won't Save Our Cities from Rising Seas."

7. Wikipedia, s.v. "Seawall," last modified July 20, 2019, 12:21, https://en.wikipedia.org/wiki/Seawall.

8. Bennington-Castro, "Walls Won't Save Our Cities from Rising Seas."

9. Bennington-Castro, "Walls Won't Save Our Cities from Rising Seas."

10. R. K. Gittman et al., "Marshes with and without Sills Protect Estuarine Shorelines from Erosion Better Than Bulkheads during a Category 1 Hurricane," *Ocean & Coastal Management* 102 (December 2014): 94–102, https://doi.org/10.1016/j.ocecoaman.2014.09.016.

11. Office of Watersheds, Philadelphia Water Department, "A Homeowner's Guide to Stormwater Management," Philly Watersheds, January 2006, http://www.phillywatersheds.org/doc/Homeowners_Guide_Stormwater_Management.pdf (archived September 2019).

12. Mark Clark and Glenn Acomb, "Bioretention Basins/Rain Gardens," University of Florida IFAS Extension, 2008, http://buildgreen.ufl.edu/Fact_sheet_Bioretention_Basins_Rain_Gardens.pdf.

13. Wikipedia, s.v. "Bioswale," last modified, May 8, 2019, 20:15, https://en.wikipedia.org/wiki/Bioswale.

14. "Storm Water Run Off: Green Infrastructure Stemming the Flow in Cities," Living Roofs, https://livingroofs.org/storm-water-run-off.

15. "Topmix Permeable," Tarmac, http://www.tarmac.com/solutions/readymix/topmix-permeable.

16. Gittman et al., "Marshes with and without Sills Protect Estuarine Shorelines from Erosion Better."

17. Michael Le Page, "Parts of San Francisco Are Sinking Faster Than the Sea Is Rising," *New Scientist*, March 7, 2018, https://www.newscientist.com/article/2163092-parts-of-san-francisco-are-sinking-faster-than-the-sea-is-rising.

18. Jennifer Gray, "Higher Seas to Flood Dozens of US Cities, Study Says; Is Yours One of Them?" CNN, July 31, 2017, https://www.cnn.com/2017/07/12/us/weather-cities-inundated-climate-change/index.html.

19. Jessica Corbett, "'When Rising Seas Hit Home': Hundreds of Towns Threatened by 2100," Common Dreams, July 12, 2017, https://www.commondreams.org/news/2017/07/12/when-rising-seas-hit-home-hundreds-towns-threatened-2100.

20. Julie Cart, "42,000 Homes in California Will Be under Water Due to Rising Seas, Researchers Project," *Desert Sun*, May 2, 2017, https://www.desertsun.com/story/news/environment/2017/05/02/california-submerging-

rising-seas-claiming-its-famed-coast-faster-than-scientists-imagined/
307228001.

21. Marianne Lavelle, " Americans in Danger from Rising Seas Could
Triple," *National Geographic*, March 14, 2016, https://www.
nationalgeographic.com/news/2016/03/160314-rising-seas-US-climate-
flooding-florida.

22. National Oceanic and Atmospheric Administration, "Economics and
Demographics," NOAA, https://coast.noaa.gov/states/fast-facts/economics-and-
demographics.html.

23. Nathan Rott, "'Retreat' Is Not an Option as a California Beach Town
Plans for Rising Seas," NPR, December 4, 2018, https://www.npr.org/2018/12/
04/672285546/retreat-is-not-an-option-as-a-california-beach-town-plans-for-
rising-seas.

24. Elizabeth Dunbar, "The 1997 Red River Flood: What Happened?"
MPR News, April 17, 2017, https://www.mprnews.org/story/2017/04/17/1997-
red-river-flood-what-happened.

25. Jerry Yudelson, "Mitigating Climate Change Damages: Managed Re-
treat?" *Reinventing Green Building* (blog), December 13, 2018, https://www.
reinventinggreenbuilding.com/news/2018/12/13/
poqzz4klrhhv2dutlljr3snhjrz8mg.

26. Earth Science Communication Team, NASA Jet Propulsion Laboratory,
"Climate Change: How Do We Know?" NASA, last modified July 31, 2019,
https://climate.nasa.gov/evidence.

27. Rebecca Lindsey and LuAnn Dahlman, "Climate Change: Global Temperature," Climate.gov, https://www.climate.gov/news-features/understanding-
climate/climate-change-global-temperature.

28. US Environmental Protection Agency, "Reduce Urban Heat Island Effect," EPA, December 12, 2017, https://www.epa.gov/green-infrastructure/
reduce-urban-heat-island-effect.

29. Christina Capatides, "Los Angeles Is Painting Some of Its Streets White
and the Reasons Why Are Pretty Cool," CBS News, April 9, 2018, https://www.
cbsnews.com/news/los-angeles-is-painting-some-of-its-streets-white-and-the-
reasons-why-are-pretty-cool.

30. Bennington-Castro, "Walls Won't Save Our Cities from Rising Seas."

31. Associated Press, "Artificial Islands off New York and New Jersey Proposed as Storm Protection," *The Guardian*, March 29, 2014, https://www.
theguardian.com/world/2014/mar/29/new-york-artificial-islands-new-jersey-
sandy.

32. Jamie Condliffe, "As the Climate Changes, NYC Is Preparing for Massive Floods," *Business Insider*, January 30, 2017, https://www.businessinsider.
com/climate-change-nyc-flooding-infrastructure-2017-1.

33. William Harris, "10 Technologies That Help Buildings Resist Earthquakes," HowStuffWorks, September 17, 2013, https://science.howstuffworks.
com/innovation/science-questions/10-technologies-that-help-buildings-resist-
earthquakes.htm.

34. Joseph Mason et al., eds., "Seismic Wave Motion," Oxford University Press, http://global.oup.com/us/companion.websites/9780190246860/stu/unit8/animation_quiz/seismic.

35. Cacciola, "Our New Anti-Earthquake Technology Could Protect Cities from Destruction," The Conversation, July 2, 2015, http://theconversation.com/our-new-anti-earthquake-technology-could-protect-cities-from-destruction-44028.

36. Harris, "10 Technologies That Help Buildings Resist Earthquakes."

37. Rithika Ravishankar et al., "Earthquake Resistance in Buildings," Slide-Share, March 26, 2016, https://www.slideshare.net/rithikarockingravishankar/earthquake-resistance-in-buildings.

38. William Harris, "How Car Suspensions Work," HowStuffWorks, May 11, 2005, https://auto.howstuffworks.com/car-suspension2.htm.

39. Harris, "10 Technologies That Help Buildings Resist Earthquakes."

40. Matthew R. Eatherton et al., "Design Concepts for Controlled Rocking of Self-Centering Steel-Braced Frames," *Journal of Structural Engineering* 140, no. 11 (November 2014), https://www.researchgate.net/publication/277609124_Design_Concepts_for_Controlled_Rocking_of_Self-Centering_Steel-Braced_Frames.

41. Harris, "10 Technologies That Help Buildings Resist Earthquakes."

42. Harris, "10 Technologies That Help Buildings Resist Earthquakes."

43. Harris, "10 Technologies That Help Buildings Resist Earthquakes."

44. R. Davoudi et al., "Application of Shape Memory Alloy to Seismic Design of Multi-Column Bridge Bents Using Controlled Rocking Approach," IIT Kanpur, September 2012, http://www.iitk.ac.in/nicee/wcee/article/WCEE2012_3228.pdf; Bipin Shrestha et al., "Performance-Based Seismic Assessment of Superelastic Shape Memory Alloy-Reinforced Bridge Piers Considering Residual Deformations," *Journal of Earthquake Engineering* 21, no. 7, August 3, 2016, https://www.tandfonline.com/doi/full/10.1080/13632469.2016.1190798?src=recsys&.

45. Wikipedia, s.v. "Carbon Fiber Reinforced Polymer," last modified June 4, 2019, 12:21, https://en.wikipedia.org/wiki/Carbon_fiber_reinforced_polymer; Martin Alberto Masuelli, "Introduction of Fibre-Reinforced Polymers: Polymers and Composites: Concepts, Properties and Processes," in *Fibre-Reinforced Polymers: The Technology Applied for Concrete Repair* (Rijeka, Croatia: Intech, 2013), https://www.intechopen.com/books/fiber-reinforced-polymers-the-technology-applied-for-concrete-repair/introduction-of-fibre-reinforced-polymers-polymers-and-composites-concepts-properties-and-processes.

46. Denise Chow, "Mussel Strength: How Mussels Cling to Surfaces," Live Science, July 23, 2013, https://www.livescience.com/38375-how-mussels-cling-to-surfaces.html.

47. Harris, "10 Technologies That Help Buildings Resist Earthquakes."

48. Lisa Zyga, "Scientists Breed Goats That Produce Spider Silk," Phys.org, May 31, 2010, https://phys.org/news/2010-05-scientists-goats-spider-silk.html.

49. Harris, "10 Technologies That Help Buildings Resist Earthquakes."

50. Victor Tangermann, "When an Earthquake Strikes, Crawl inside This Egg-Shaped Safety Pod," Futurism, December 7, 2018, https://futurism.com/the-byte/egg-pods-humans-save-life-earthquake.

51. Jason Meyers, "Moon Landing Anniversary: Weather on the Moon," NBC26, July 20, 2016, https://www.nbc26.com/storm-shield/storm-shield-featured/moon-landing-anniversary-weather-on-the-moon.

52. Kathryn Mersmann, "The Fact and Fiction of Martian Dust Storms," NASA, September 18, 2015, https://mars.nasa.gov/news/1854/the-fact-and-fiction-of-martian-dust-storm.

53. Tim Sharp, "Mars' Atmosphere: Composition, Climate & Weather," Space, September 11, 2017, https://www.space.com/16903-mars-atmosphere-climate-weather.html.

54. Sharp, "Mars' Atmosphere."

55. Mersmann, "The Fact and Fiction of Martian Dust Storms."

56. Matt Williams, "What Is the Weather Like on Venus?" Universe Today, December 3, 2016, https://www.universetoday.com/36721/weather-on-venus.

57. Mike Wall, "Incredible Technology: NASA's Wild Airship Idea for Cloud Cities on Venus," Space, April 20, 2015, https://www.space.com/29140-venus-airship-cloud-cities-incredible-technology.html.

58. Wikipedia, s.v. "Instrumental Temperature Record," last modified August 1, 2019, 8:35, https://en.wikipedia.org/wiki/Instrumental_temperature_record.

59. Bruce Mapstone, "Remind Me Again: How Does Climate Change Work?" The Conversation, March 28, 2011, https://theconversation.com/remind-me-again-how-does-climate-change-work-46.

60. Ker Than, "The 100-Year Forecast: More Extreme Weather," Live Science, October 19, 2006, https://www.livescience.com/4231-100-year-forecast-extreme-weather.

61. Uma Sharma and Alyssa Pagano, "Here's How We'll Control the Weather in 100 Years," Business Insider, July 19, 2019, https://www.businessinsider.com/controlling-weather-in-future-2018-5.

62. "The Causes of Climate Change," NASA, https://climate.nasa.gov/causes.

63. Stuart Fox, "The Top Ten Greenhouse Gases," Popular Science, March 17, 2009, https://www.popsci.com/environment/article/2009-03/top-ten-greenhouse-gases.

INDEX

ABOUT THE AUTHOR

Jeff Dondero has a diverse background and much experience in writing, ranging from B2B articles, books, hard news, and interviews to feature writing to internet content. After studying business and law at the University of San Francisco, and communications, broadcast arts, English, and journalism at San Francisco State University, he began his career as stringer and freelancer for the *San Francisco Examiner*. He worked as a reporter, editor, and contributor for several newspapers and magazines in the San Francisco Bay area, was the entertainment editor for the *Marin Independent Journal*, wrote for KTVU-TV, toiled in a trade magazine mill, and cocreated a website dedicated to the technology of sustainable construction industries, the *Green Building Digest*.

He was invited to be writer-in-residence at an art colony in Rancho Linda Vista, Arizona, in 2014, where he wrote a slim volume of poetry and researched material for upcoming books. He continues to expand his national readership with books, social media, various writers' blogs, websites, and radio and television appearances.

His previous books include *The Energy Wise Home*, *The Energy Wise Workplace*, and *Throwaway Nation*, all published by Rowman & Littlefield. Other books include *Do You Want to Survive (24) Natural Disasters*, *The Marin Companion*, and *Brutal Beauty*, a collection of poems written as a writer-in-residence in the Rancho Linda Vista art colony.

Jeff Dondero lives and works in Rohnert Park, Sonoma County, California.